Foreword

Much has been talked about the revival of the tenement form in Scotland. Indeed Glasgow's regeneration can be traced to the rise of tenement rehabilitation in the early 1970s. The take-up of improvement and repair grants, together with the rise of the Housing Association movement, have brought about considerable improvements to the older tenemental stock. However, one problem remains. How are we to maintain these tenements if they are to last into the next century? The problems of joint ownership remain, and there are few signs that tenement owners truly realise the need to form stair associations or even inspect and maintain the common property at regular intervals.

We hope this handbook will help those who wish to set up their own stair association, understand their tenement more fully and become more aquainted with the various legal and management issues which can beset the tenement owner.

John Gilbert
ASSIST Architects Ltd
4 Sept 1992

From 1972 - 1983, ASSIST operated as an architectural Practice and Research Unit of the Department of Architecture and Building Science of Strathclyde University.

Its early work in Govan on public participation in tenement improvement lead directly to the organisation of Glasgow's Rehabilitation Programme around local community-based Housing Associations. This symbolic reversal of decline formed the basis of the popular regeneration of Glasgow's housing and its poor image.

Our early involvement in housing grew to encompass Common Repairs Projects and Environmental Improvement landscaping work in tenemental areas, working closely with owners and residents.

The problems of urban areas highlighted the need for solutions outwith housing to complete area renewal and ASSIST has been involved in many Community Projects to meet these needs.

With this solid background and a good track record, ASSIST became involved in the wider field of employment, commerce and leisure, usually from a community initiative. Several of these projects (Govan Workspace, Elderpark Workspace, The Briggait) received awards in their field.

In 1983, ASSIST broke away from the University and started working independently as a co-operative architectural practice, ASSIST ARCHITECTS LIMITED, taking over all current jobs of the ASSIST Unit.

The last decade has seen a change in work type towards New-build Housing and refurbishment of relatively new post-war housing stock. This has encompassed a challenging variety of housing types ranging from the low-rise system-built blocks to industrial multi-storey towers. By continuing the principles of participation and involvement, we have provided new and improved housing which match the user's needs, builds on the process of area regeneration and adds to the urban framework of our towns and cities.

ASSIST is also pleased to have been involved in work outwith the main conurbation of Glasgow and has completed a town centre study in Stirling, building upgrading in Edinburgh, building studies in Lanarkshire and Integrated Housing in Ayrshire.

ASSIST intends to continue to provide a quality architectural service, by involvement and participation, to its clients on a wide range of projects. We aim to develop our skills, interests and awareness to improve our service, particularly in the aspects of energy efficiency and conservation on a broad scale.

Acknowledgements

Thanks are due to the R.I.B.A. who provided a grant to ASSIST Architects to cover most of the initial research. Many other individuals gratefully gave help throughout the long period of production.

ASSIST Architects and the Authors would also like to thank the following:

Jim Murdoch from the Department of Public Law, Glasgow University;

Norman Wagstaff from the Computing Department, University of Strathclyde;

Staff at the Mitchell Library; Strathclyde Fire Brigade; Strathclyde Regional Council for their help on consumer and trading standards; Glasgow District Council's Private Sector and especially Arthur Glendinning, Dennis Rodgers and John Murray; C.S. McAllister; Peter Robinson from the Scottish Courts Administration; Colin Shaw FRICS; Sebastian Tombs of the RIAS.

We would also acknowledge the Edinburgh New Town Conservation Committee's publication 'The Care and Conservation of Georgian Houses'.

Allison Morris and Josephine Mitchell who spent many hours trying to create a manuscript from our many notes, and to Kate Cornwall-Jones for her grammatical checks.

Thanks are also due to the many different self-factoring groups in the Govanhill and Mount Florida areas of Glasgow who reported their problems and successes.

Produced by:

ASSIST Architects Ltd.

Station Terrace, 100 Kerr Street, Bridgeton, Glasgow G40 2PR.
Tel: 041 554 0505. Fax: 041 554 6112.

Published by:

RIAS, 15 Rutland Square, Edinburgh EH1 2BE.

© ASSIST Architects Ltd.

1SBN 187319014X

Illustrations:

John Gilbert.

Duncan Roberts of City Design Co-operative Ltd.

Cartoons:

Marcus Gray of City Design Co-operative Ltd.

Indexer:

Oula Jones.

Typesetting:

Typescale Ltd.

Printed in Great Britain at the Alden Press, Oxford

Contents

Managing Your Tenement

Organising the owners

Tenements are often taken for granted when in fact they are large complex structures, which have to be cared for in much the same way as, for example, a car does. The important point to note is that, jointly, those living in tenements will fair much better if:

> *they can work together and they take responsibility for the building themselves.*

Otherwise there's always the risk of losing control (either by neglect or by intervention from the Local Authority).

This chapter deals with the task of managing your tenement and working with other owners and occupiers. You may employ a special property agent known as a factor or you may prefer to do this yourself. Whichever you choose, you will need to be well organised and to work with your co-owners in a fairly business-like way. This generally means having a meeting.

Using a Factor

In the west of Scotland you can hire a professional property manager called a **factor** to manage your tenement. The factor's main duties are:-

> *Dealing with repairs and tradesmen.*
>
> *Collecting everyone's share of tenement common repair bills.*
>
> *Payment of common insurance and feus.*
>
> *Handling rents and repairs for tenanted property.*

Most factors charge a management fee of about £40 per year per owner. In addition you will pay for repairs, insurance, feus and VAT. Many factors charge extra for writing letters, arranging larger repairs, or dealing with grants.

If you want to hire a factor you can ring round names found in the Yellow Pages under **Estate Agents, Estate Managers** and **Property Management.** There is no specific listing for factors. You can also contact the *Property Managers Association Scotland Ltd., 2 Blythswood Square, Glasgow G2 4AD Tel: 041 248 4672 (Contact Mr J Millar).*

Most factors ask people to sign a contract which details their duties and responsibilities *(we explain this contract in plain English in appendix 1).*

If you don't like the contract you can have a meeting with the factor to discuss how you would like the factor to work in your tenement. However, any factor's contract must comply with the arrangements specified in your Title Deeds. Therefore you should check your **Title Deeds** before drawing up a new contract *(see Title Deeds on page 5).* If you are looking for a better service than that set out in the contract, you will pay more for it.

It is important that you understand that the factor is your Agent. The dictionary says that an agent is *"someone authorised or delegated to undertake work on behalf of another".* The *'on behalf'* is important as it puts a legal obligation on you as well as on the agent. If you have a factor, you should read the section called **The Law of Agency** on page 2. It is very important that you realise that taking on a factor does not relieve you of all responsibility.

There was a very important court case in 1872. *A tenement was owned by one landlord. One of the banisters was*

Organising a Close Meeting

Some Tips
Decide who is to be invited.

Do you want the factor or architect to come? If you are talking about repairs, be sure to ask any landlords there might be. If you are talking about stair cleaning then you should ask tenants to attend.

Pick a time when everyone will come. Don't pick Cup Final night and time the meeting to miss Coronation Street and other popular TV programmes. Sunday afternoon might be good time if everyone is out in the evenings.

Do your homework and remind others to do theirs. Write out a list of the topics you want to talk about and give your neighbours a copy.

Help your neighbours get to the meeting. You might need to help an old person up the stairs or get baby-sitters. Make the meeting cosy. Beg, borrow or steal enough chairs and cups to go round.

At the meeting stick to talking about one thing at a time and reach a decision on one item before going on to the next item. Decide what you want to get out of the meeting and how you are going to get it. If people start to

go on at length, remind them politely of how many other things you've got to talk about.

Decide if you want to have someone chair the meeting or not.

Discuss how decisions are to be taken - by full agreement of all owners or by a majority decision.

Try to get everyone involved in the work. Someone who works in an office may be able to make photocopies, while someone who is home all day can keep an eye on workmen and hold keys.

During the meeting, get someone to take notes. At the end of the meeting go through the notes and remind people what has been agreed and what they are going to do.

Discuss whether or when you need to have another meeting.

If you cannot agree about what to do, you can ask an outsider to come and help resolve your differences. The outsider could be the factor, someone from the Residents Association or Community Council, an architect or another consultant.

Let people who weren't at the meeting know what has happened. Try to find out why people weren't there.

missing and this was reported to the factor by one of the tenants. Nothing was done about it and a little girl fell through the banisters and was killed. In the court case, the owner and not the factor was held responsible.

As an owner, you are jointly responsible for the upkeep of the common property. If anyone is hurt, e.g. by a falling chimneyhead, you will be held responsible in law. However, if you report a defect to the factor, and he fails to act, you may be able to take action against the factor in the event of an accident.

To take this a bit further, you cannot expect the factor to be any more concerned about the state of the property than you are. You cannot expect the factor to tell you what to do, or give advice, for example, about what grants are available, because the factor is an **agent** not an adviser. You should regard the factor as someone you have taken on to manage and administer the property and not as someone you have retained as a technical consultant. Do not expect the factor to push reluctant owners into agreeing to repairs.

Consultation

There are a number of common complaints made about factoring, one of which is lack of consultation. You cannot do much about this because of the factor's 'implied authority' as an agent. If you have been wrongly billed, you should **complain in writing** to the factor. You cannot refuse to pay the whole bill, only the part that you are disputing. Factors are getting into the habit of taking non-payers to court. If you do not handle the matter properly you could lose out. If the factor insists on taking out a summons against you despite your complaints, be sure to go to court and defend yourself *(see Taking Legal Action on page 37).* If you cave in at the last minute you will end up paying the whole bill plus legal expenses. If you put all communications in writing, and keep copies of all letters, you will be able to use these as evidence in court.

Standard of Work

Another source of complaint is the standard of repair work organised by the factor and carried out by a tradesman.

If any repair is unsatisfactory be sure to complain to the factor as soon as possible. You will still be liable to pay the bill. Unfortunately, the later you complain about poor work, the easier it is for the tradesmen to make excuses and to say that it has been the fault of the weather or something else. The factor will also have paid the tradesman and won't be able to use the fact that the bill hasn't been paid as a lever to ask them to improve the repair.

THE FACTOR

If the flat owners in your tenement have a series of complaints against the factor, you should think about getting better organised. You may just want to discuss these complaints among yourselves and then give your factor additional instructions. For instance, you can tell the factor you want all tradesmen to report to a certain owner at the start and finish of every repair job. If you're still not happy, think about doing it yourself. Information about how to manage your own stair is given below. If things go seriously wrong, you may be able to sue for **Negligence** and **Breach of Contract.** (There is further useful reading in *Scottish Consumer Council (1984) Under One Roof. Edinburgh: HMSO).*

Stair Associations

If you do not have a factor it is a good idea to have a formal agreement with other flat owners about how your tenement should be managed. You can do this by setting up a **Stair Association** with written rules and a bank account.

Use the rules on page 3 as a guide to drawing up your own set of rules. Discuss them fully with other owners and check that they do not conflict with what your **title deeds** say. Even if they do conflict, there is no problem if every owner agrees and signs the rules.

Having a well managed stair will be a positive point when it comes to selling a flat in the tenement. New owners should be asked to join the Stair Association and sign the rules.

Obviously you will want to make sure

The Law of Agency

There is a specific law of agency which governs what you and your agent (e.g. the factor) should do.

The Agent:- Can expect to be paid for doing what is legally your responsibility.

Should take care of the property according to the general standards of the profession.

Should carry out your instructions. If there are no instructions, they should act to the best of their judgement.

Has implied authority to act in accordance with the general principles of their trade or profession.

Should not permit any conflict between their own personal interests and yours.

You:- Can authorise the agent to do anything you can legally do.

If the agent is acting within their authority you are responsible for all the outlays incurred by the agent in the course of their duty. So, if a tradesman is correctly called in by the factor to do work for you, you are responsible for the bill, not the factor.

If you give the agent specific instructions, they must keep to these. If they do not you can hold them responsible for the bill. So, if you tell the factor not to employ a certain firm, and the factor ignores your instruction and goes ahead, you can tell the factor to foot the bill.

If the agent takes on a contract for you, you alone, and not the factor, can sue or be sued as a group of owners.

For more information on the Law of Agency see *Enid A Marshall (1980) Scottish Cases on Agency - Green: Edinburgh.*

that all owners take their share of jobs. Some owners may be reluctant to have a go at something they have never tried before. You must make sure that they do try, perhaps with someone there to help out. If everything is be left to one person and that person then leaves, the whole Stair Association could fold. If someone really cannot help with organising repairs then they could clean the stairs, keep the

from interest paid on personal accounts. If your Stair Association has a clause in its constitution which says that all interest accrued on the account will be spent commonly and not distributed to individuals, the bank may not deduct tax from you. Ask before you open the account and shop around if necessary.

Dealing with Interest Earned on Accounts

You cannot divide up interest among individual owners if you want to avoid paying tax on interest. Use the money towards repairs or even towards a stair celebration such as Bonfire Night or a Christmas Party. If someone leaves, you will need to give them back what they have contributed (without the interest). Alternatively you may decide that owners who sell should leave their fund in the account and pass the benefit on to the new owner. The money held in the account would be set off against the sale price.

A Stair Association Meeting

garden or just be assigned to keep the keys if workmen need to be let in.

The **model rules** here suggest that one of the owners should act as factor. This can be useful when dealing with tradesmen who all understand what a factor does but who may be confused about what a Stair Association does.

You should make some provision for emergency repairs and make sure everyone knows what they are. You will also want to make sure that people do not organise repairs without consulting other people. Make sure that you are ready for an emergency and that everyone else knows what to do.

Opening a Bank Account

All banks have accounts which are specifically designed for groups. You will need to get a form signed by **at least** two people. Two of these signatures will be needed for each cheque for withdrawal. You may want a **Current Account.** With this kind of account you get a cheque book.

Banks and Building Societies deduct tax

Keeping Records

You do not need to be an expert book keeper to keep accounts. The main thing to is to keep a book and write every payment made or received into this book. Keep copies of every bill. It's a good idea to have a small petty cash fund to cover the cost of stamps, phone calls, etc.

Keeping in the Black

There are two main ways of making sure you have always got enough money in the stair account to cover the cost of repair:-

> *Get everyone to pay in a float and top up the account every time you get a repair bill. You should aim to keep at least £300 in the bank. When you know in advance that you are getting work done, get people to pay in advance.*

> *You can get owners to pay a small amount regularly. If owners won't or can't contribute to the scheme, you will just need to send them a bill for their share.*

The Law of The Tenement

Your tenement is governed by common law rules of common interest and common property but these rules are superseded by your Title Deeds.

Common Property

Common property refers to parts of property which belong to everybody and which everybody must pay for jointly. The definition of common property is to be found in your Title Deeds. *Illustration 2:1 overleaf shows what is normally considered to be common property.* Check your own **Title Deeds** in case there are minor differences. In the exceptional case that there are no title deeds or the deeds do not cover a specific part of the building, then the **common law** applies. Under common law the roof belongs to the top flat owners, the ground and gardens to the ground floor owners and each owner owns the walls around their flat to the half way mark with the next flat. Under common law only the stairs and passageway are in fact common. It is important to remember the common law definition is not in common use!

Note: at the time of writing the Scottish Law Commission is finalising its report on a review of the Law of the Tenement, which could alter these definitions.

LOOKS LIKE THAT TENEMENTS BECOME A REAL BURDEN!

Common Interest

Common interest protects your rights over the parts of the tenement which are not in your individual ownership - it is a mutual obligation which each tenement flat owner has towards other owners in the same tenement.
For instance, every tenement owner has a common interest in the roof being maintained, the walls not being knocked down and the backcourt not being turned into a junk yard.
Common interest could give you grounds to demand payment for common repairs if there is nothing expressly contained in the Title Deeds. *See a lawyer first.*

Title Deeds

Most flats have **Title Deeds**. If you have a mortgage, the deeds are kept by the lender. When you buy a flat, your solicitor will look at the Title Deeds before sending them to one mortgage lender. Your solicitor should also advise you as to what the deeds mean in terms of common repairs, your share of costs at the time of purchase. If your Title Deeds are now with the Building Society you may have to pay a few pounds to get a copy. Copies of registered deeds are also available from the *Register of Sasines in Edinburgh, Registers of Scotland, Meadowbank House, 153 London Road, Edinburgh, EH8 7AW. Tel: 031 659 6111.*

Where Scots Law comes from

Land in cities has always been densely built up with one family living on top of the other. Scottish law is based on Roman law and similar to the law of other countries in many respects. The Romans had tenements and they had a concept of law - 'servitude' - which allowed each building to be owned by a number of individual owners. Servitude is a burden on a property obliging the owner to allow another person an easement. In the case of the tenement, the **servitude onus** *ferendi* allowed one owner to build on top of another and have a right of support. (See Common Interest in this section).

The Scots developed this law through **Burgh Police Acts** and **Dean of Guild Courts.** These were local by-laws and local courts. Over the period 1800-1900, these rules became tighter and tighter, laying down minimum room sizes, wall thicknesses, ceiling heights and even the actual number of people allowed to occupy each flat. Many of these rules were developed in the interests of public health and attempted to control ruthless property developers, who built housing to let when many people were moving to the city from rural areas because of clearances and other economic changes. Later, the builders themselves imposed strict Deeds of Conditions in **Feus** and **Title Deeds** in an attempt to persuade prospective buyers that their tenements would remain exclusive properties of high amenity.

Title Deeds are now being computerised and registered. This is happening district by district. If your area has been registered, you will be given the old deeds to keep.

Missing Title Deeds?

Firstly, check with the **Register of Sasines** (see above). If there are Title Deeds, they will certainly have a copy. If there is nothing in the Title Deeds on a particular subject then the Common Law will apply in that instance. This says:-

The Roof: where there are 2 or more top flats, each top flat owner owns the roof directly above their individual flat. The section of roof above the common stair is commonly owned. The top floor flat owner also owns the gutters and the length of downpipe going down to the next floor level. The chimney pots are individually owned but repairs to the stack would need to be paid for by all those who had chimney vents in it.

The Plumbing: belongs to the flat wall it is attached to.

The Walls around the common stair are common as far as the halfway mark with any flat wall.

The Floor Joists are owned individually up to the halfway mark with the next flat.

The Ground under the tenement (known technically as the Solum) belongs to the owner of the ground

floor flat immediately above. The ground below the common stairs is common property.

This definition of the common law makes **common interest** very important. It allows an owner the right to use pipes or chimney vents running through another owner's flat and allows one owner the right to rely on another owner to keep their property in decent repair.

What's in the Title Deeds

The deeds are probably a collection of papers containing the original **Deeds** and subsequent **Dispositions** showing later sales of the house. There may be other papers, redeeming the feu, making changes etc. The information you want will be in the original **Deed of Conditions**. The deed will divide itself into sections as follows:-

Details of original disposition - who sold what to whom and for how much.

Details of property, e.g. flat position, land boundaries, which county the property lies in.

Details of common property, feu payments etc.

How you must come to agreement with other owners about repairs, when you must consult other owners and how much each owner pays for repairs.
Insurance.
The factor's duties and arrangements

for using them.
Your obligations to other owners.

Details of arbitration procedures.

What happens if people do not pay.

A final declaration that these are real burdens which will not change when a flat is sold.

Your title deeds may have some useful clauses prohibiting

your neighbours from doing certain things and obliging them to give access for repairs etc.

Use the list of legal phrases opposite to help you make sense of your Title Deeds.

Changing the Title Deeds

You can still make changes to your Title Deeds. If everyone in the tenement agrees to the changes, you can ignore the Title Deed. However this can create problems when one owner sells and the new owner doesn't like the changed arrangements. It is possible to ask the **Lands Tribunal** to make some changes in the Title Deeds. You cannot change things in the Title Deeds referring to feu duty or other money paid to people outside the tenement. Your solicitor can advise you on whether a particular change is possible.

It is fairly cheap to make an application to the Lands Tribunal but you would be advised to have legal assistance and this could be expensive. The current cost of an application varies between £60 and £100 depending on the type of

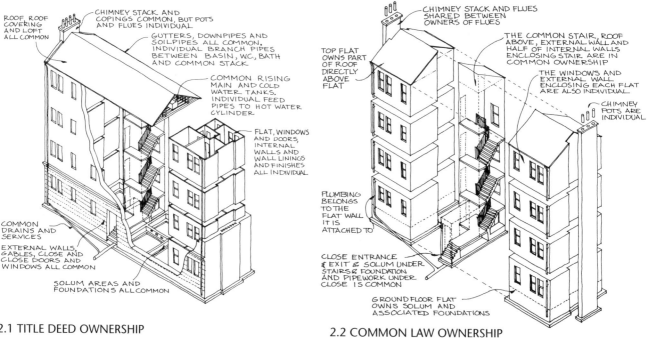

2.1 TITLE DEED OWNERSHIP

2.2 COMMON LAW OWNERSHIP

Words and Phrases Used in Title Deeds

Ad Longum - at length.

Arbitrator - an outsider called in to make a judgement between two disagreeing people, often a lawyer.

Appurtenances - accessories or rights belonging to a property.

Assign - to make over or transfer.

Assignee - the person something is transferred to.

Assessed Rental - rateable value.

Bond - a mortgage or a written obligation to pay money.

Burden - see Real Burden.

Convene - to call a meeting.

Cumulo as *Cumulo Assessed Rental* - all these amounts added together.

Decree - a legal order.

Delineated - outlined.

Dispone - to transfer a property to another person.

Disposition - a legal document showing that the property has been transferred to somebody else.

Easement - a right which a person may have over another person's land or property, such as the right to access a certain place or run a pipe under a neighbour's tenement.

Effect - to do something.

Effeiring - belonging to.

Execute - to carry out a task.

Egress - way out.

Exigible - the amount which must be paid.

Feu - Tenure of land in perpetuity in return for continuing annual payment of a fixed sum (feuduty) to the owner of the land (feu superior). Most feus now redeemed (bought out) and feus likely to be abolished shortly.

Free Ish and Entry - the right to come and go.

Ground Annuals - similar to Feuduties (developed to get round the old restrictions against subfeuing).

Heritable Property - a building or piece of land that can be freely bought and sold. It also includes rights over, or connected with, land and buildings (e.g.the right of a tenant).

Instrument - piece of writing containing a contract.

Investitures - a letter etc., showing authority to do something.

Irritate - to make something null and void.

Ish - to be able to go out.

Lien - a right to hold someone's property until a debt is paid.

Mean - a line dividing an uneven thing into two equal parts.

Nominee - someone who is named to carry out a task on another person's behalf and with their authority.

Pertinances - something which belongs to the property or a person.

Primo - first item or person.

Proprietor/Proprietrix - the owner.

Quarto - fourth item or person.

Quoad - as far as.

Real Conditions - obligations affecting the use and enjoyment of land, usually imposed by superiors when referring to buildings to be created on such land to define the use of the land (and hence buildings).

Real Burdens - a restriction or duty placed on heritable property or the owner of such property which can be transferred from one owner to the next.

Real Property - used in English legal terms to mean property in the form of land and buildings. In Scotland, Heritable Property is the nearest equivalent term.

Resolutive Clauses - clauses which define or sort out an issue or property.

Secundo - second item or person.

Servitude - an obligation that runs with the property which obliges an owner to allow other people to do certain things.

Solum - the ground the property stands on.

Steading - farmed or cultivated land.

Tertio - third item or thing.

Transmission - a transfer of property.

Videlict - namely (often shortened to viz).

Warrandice - a guarantee.

Writ - a legal document.

---

application. If there are any objections made to your application a hearing will be required. This will cost the applicant £120 per day.

The Lands Tribunal has the power to change land obligations which are:

Unreasonable or inappropriate because of changes in the neighbourhood.

Unduly burdensome compared with the benefits.

Where the existence of the obligation prevents some reasonable use of the land.

Your share of common repairs is a land obligation but the problem is that while the Lands Tribunal can say one owner should pay less, they have not got the power to say that another owner should pay more.

A case went to the Lands Tribunal concerning common repairs. The case arose because the original owner of a whole tenement sold off flats and a shop , but kept on another shop for many years. The Title Deeds said that one shop had to pay 40% of repair costs while the owner's remaining shop only had to pay 3%. The owner of the shop paying 40% decided to go to the Lands Tribunal. The case was unsuccessful because the other

owner could not be forced to pay more. (1)

Contact

The Lands Tribunal, 1 Grosvenor Crescent, Edinburgh EH12 5ER. Tel: 031-225 7996

Legal References

(1) Section 1. (iii) Conveyancing & Feudal Reform (Scotland) Act 1970.

Further Reading

Connell I: Law affecting building operations and Architects and Builders Contracts. 1903. (Contains chapter on law of tenement).

McAllister, A, and Guthrie, T: Scottish Property Law. 1992. Butterworths.

Common Repairs

PAGE8

1. Chimney stacks. But may be mutual with next close.

2. Chimney vents or flues. Items such as dropping a flue liner to allow a gas fire to be installed are individual but the structure of the flues should be common.

3. Roof. Includes flashings, ridges, roof covering and access rooflights.

4. Roof timbers.

5. External walls.

6. Mutual gable walls. These are owned in common with the next door close.
All owners in your close and all owners in the next door close should pay for repairs to these walls.

7. Internal walls. Individual but if the wall is dividing two flats then costs should be shared between owners of these two flats.

8. Close walls. The flat owner owns the half facing into the flat but the half facing the close is common. The flat owner might have to pay half the cost of repairs.

9. Lintols.

10. Foundations.

11. Bressumer beams or any other beam in the external walls.

12. Gutters.

13. Downpipes.

14. Stair windows.

15. Bannisters.

16. Close stairs and access doors into and out of close.

17. Close decoration.

18. Front fence, including gates.

19. Solum.

20. Back garden.

21. Path.

22. Bin stores.

23. Clothes poles.

24. Back fence. But may be shared with next door.

25. Back court walls. But may be shared if between two closes.

26. Electricity. Responsibility of Scottish Power as far as meter, then individual.

27 Gas. Responsibility of Gas Board as far as individual flat walls - then individual

28. Stair lighting. Responsibility of the local Regional Council, but any improvements are a common repair.

29. Water tanks, unless they are placed inside the individual's flat and only serve that owner.

30. Mains supply pipes. Common until the branch pipe to individual flat where they become individual. *(not shown on drawing)*

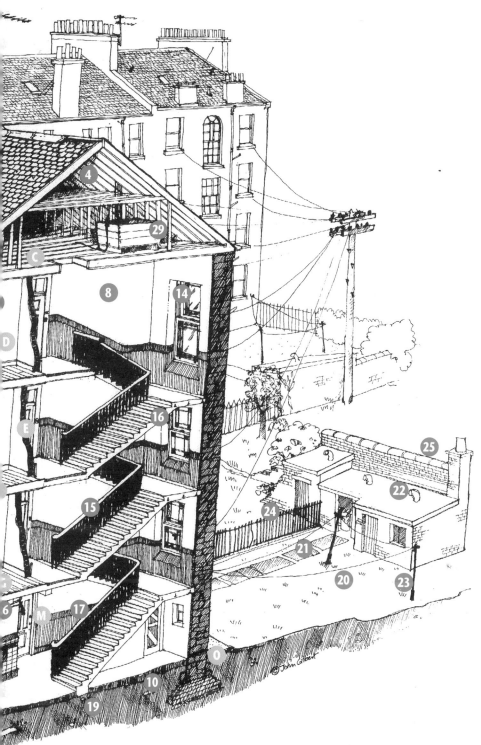

A Chimney pots and cowls.

B Chimney vents and flues.

Items such as dropping a flue liner to allow a gas fire to be installed are individual but the structure of the flues should be common.

C Roof Insulation.

D Internal walls*. Individual, but if the wall is dividing two flats then costs should be shared between owners of these two flats.

E Close walls*. The flat owner owns the half facing into the flat but the half facing the close is common. The flat owner might have to pay half the cost of repairs.

F Safe lintols.

G Floors.

H Ceilings.

I Floor joists*. Owned to halfway mark by flats above and below. If the ends have rotted away completely, then each owner would have to pay half the cost of repairs.

J Connection from flat to downpipe*.

K Mains supply pipes*. Common until the branch pipe to the individual flat where they become individual.
(not shown on drawing)

L Flat windows, including roof windows and dormers which light the flat.

M Front doors to flats.

N Front gardens.

O Damp proof course*.

*If you are carrying out a big common repair scheme then these items will be treated as common and not individual.

2.3 THIS DRAWING SHOWS HOW THE MAJORITY OF TITLE DEEDS DIVIDE UP REPAIR RESPONSIBILITIES

Organising Common Repairs

This chapter tells you how to organise a common repair with your neighbours. Most of it also applies to repairs organised by factors.

What is a Common Repair?

A common repair is any repair to the common property. Your Title Deeds may be more specific about common repairs.

There is nothing to stop you agreeing that something should be a common repair, **providing all owners agree** *(see fig 2.1)*. There may be an advantage in leaving some things as individual repairs if that owner can get insurance to cover the loss or a much higher level of grant.

There is also an advantage in defining a common repair as something which has to be done because of a direct lack of common repairs. *For example, if a leaking roof has brought down the ceiling of the top floor flat, the repair of the ceiling should be paid for by everyone. However you will also need to clarify if the decoration that follows the repair is paid for individually or mutually.*

Dividing Up The Bill

You should check your Title Deeds to see exactly what applies to your tenement. You will probably find that one of the following applies:-

All owners pay equally

Owners pay according to Rateable Value (sometimes called Ground Rental)

Owners pay according to amount of feu duty paid

Owners pay according to a set fraction or percentage

Taking Decisions About Common Repairs

Again, check your Title Deeds. They will say:-

How you must come to an agreement, for example, whether everyone must agree or whether a majority vote is sufficient to initiate any action.

How many estimates you must get for each repair

Calculating the Rateable Value Share

The **Rateable value** is often called the 'RV' for short. You can find the rateable value of the flats in your tenement by looking up the old Valuation Roll. The Regional Council's Assessor may have copies of the valuation roll for 1989 (this was the last year the **valuation roll** was prepared).

You must be careful to find the Rateable Value not the actual amount of rates paid. Add up all the RVs to get the total for the stair. Divide the total by each individual owners RV to get their share of the cost. You will probably need a calculator!

In future, you may need to use the council tax valuations.

If someone refuses to agree, you have two courses of action:-

Compulsory Repairs Notices (see page 12).

Get the repairs done and then take any owner who won't agree to Court (see Taking Legal Action on page 37).

The repairs notice will generally be best because the council deals with the reluctant owner - not you. Once called in, the Council's officers may often require additional work to be done.

Common Repair Survey

If you are intending to appoint a builder to carry out repair work, it is adviseable to inspect the property beforehand and make a full list of the repairs needed. You could hire a firm of architects or surveyors to carry out an annual maintenance survey for you or you can do your own survey using the technical section of this manual as a guide. If you are considering using professional help read chapter 4 first.

Common Repair Trouble Spots

Dry rot.

Dry Rot is a fungus that occurs in constantly wetted but unventilated space, it feeds on timber mainly but it can travel great distances and causes a lot of damage (see chapter 18). Try to identify the cause of the problem: if it stems back to an individual repair not being done, then that individual could be held responsible for paying for the whole repair, including work in other people's flats. By the same token, a common repair not done could make everyone responsible for paying for repairs in an individual flat. The problem is that some owners will refuse to pay for what they see as other people's problems and there is very little in Title Deeds (or the common law) on this matter.

Damp proof courses (D.P.C)

A damp proof course is an impermeable layer stopping water rising up the walls from ground level. Such moisture can lead to dry rot which can affect other houses in the close (see chapter 17). The problem here is that this is often a common repair but only one owner gets the benefit in terms of the increased sale value of their flat. That owner may also be able to get an **improvement grant** for putting in a **D.P.C** and this may be available in addition to, or instead of, a **repair grant**. Very few Title Deeds mention D.P.Cs and they may either be considered an improvement or a repair to a wall.

Emergencies

In an emergency anyone can act as an *'agent of necessity'*, get the repair seen to, or, at least made safe and then get everyone else to pay up.

Not being consulted

If someone orders a common repair without getting other people's agreement you can refuse to pay if you can show that person wasn't acting in your best interest. You may need to pay some of the bill, if you have benefited from the repair.

Getting agreement

Many repair schemes fail to take off because some tenement flat owners will not agree to repairs. Some owners will

have genuine problems or fears. These can be:-

Not being able to pay.

Unable to cope with the idea of repair. Try to find out what they are worried about and explain carefully what will happen - let them have a look at this book.

Unable to cope with building work. How can you as neighbours help? For real problems try the Social Work Department or the Housing Department.

Language Problems. Much of the language used in legal documents is technical. Some important words and phrases are explained in Chapter 2. If you do not understand any word or phrase always ask other neighbours or a solicitor what they think it means. You may also need to track down some owners (see absentee owners on page 41). If an owner is on the verge of selling, try to persuade them to pass on the terms of agreement to the new owners as part of the sale terms.

Compulsory Repairs

There are a number of ways you can force someone to get repairs done:

Section 108 Notice Housing (Scotland) Act 1987

(previously called a Section 24 notice).

This must be served by the council (see the Housing, Environmental Health, Building Control or Technical Services departments). This covers serious disrepair or repairs which could lead to serious damage if not seen to. The notice can be served on a house which is in a state of serious disrepair and this is extended to include cases where the situation is likely to deteriorate rapidly or cause material damage to another house (1).

It can be served on the 'person in control of the house' meaning *'the person who would be entitled to the rent'* and includes a trustee, factor or agent (2).

Appeal must be within 21 days of the notice being served and the council can do nothing until the Appeal is decided. You may appeal against the notice and against a charge being made (3).

Section 109(1A) Housing (Scotland) Act 1987.

This allows the council to recover their expenses in serving the notice and Section 108 part 3 of the same Act allows the council to carry out the work themselves after a certain time and allows more work to be tackled if this couldn't have been foreseen at the time. In addition to a charging order, the council can also demand payment of costs by instalments.

The council can charge the person who collects the rent for the flat but can't make this person pay more than

the rent they have 'in hand' (4).

The council is obliged to make grants available if you qualify and, if you don't qualify, the council is obliged to give you a loan (5).

Section 87 Notice Civic Government (Scotland) Act 1982.

This must be served by one of the council departments above. This is a wide ranging notice covering any kind of repair.

The Notice covers any work needed to bring a building into a reasonable state of repair according to the age, type and location of the building. Binstores, back walls, paths etc. can also be covered by the notice.

Section 99 of the Civic Government Scotland Act 1982 allows the council to step in and do the work required in the notice if the owner(s) do nothing. Owner(s) will be charged the cost of the work.

Section 100 allows the council to charge interest on costs and **Section 108** allows the council to recover expenses through a charging order.

An appeal must be made by lodging a summary application within 14 days (6). The council must do nothing until the appeal is heard (7).

Anyone who tries to stop the council entering their premises to do work under Section 99 can be found guilty of an offence and fined £200 (8). If there is any difficulty getting entry to a house to do work *(i.e. because the house is empty)* the council can ask the Sheriff for an entry warrant (Section 102)

An occupier (e.g. a tenant) can do work specified in the notice and deduct money from rent (9).

A Public Nuisance Notice

This is served by the Environmental Health Department but - and this is the difference from all the other Compulsory Repairs Notices - you can force the council to serve a notice (See **Nuisance** and **Public Health**).

An Improvement Order

This can be served on a house which does not meet the **Tolerable Standard.** This order must be served by the council but the owner will be entitled to grants and loans. **(Section 88 Housing Scotland Act 1987).**

(1) Section 108, part 4 Housing (Scotland) Act 1987.

(2) Section 208, part 2 Housing (Scotland) Act 1966.

(3) Section 111 Housing (Scotland) Act 1987.

(4) Section 14, part 1 Housing (Repairs and Rents) (Scotland) Act 1954.

(5) Sections 8 and 9 Housing (Financial Provisions) (Scotland) Act 1978.

(6) Section 106 Civic Government (Scotland) Act 1982.

(7) Section 107 Civic Government (Scotland) Act 1982.

(8) Section 101 Civic Government (Scotland) Act 1982.

(9) Section 103 Civic Government (Scotland) Act 1982.

Getting It In Writing

If you are committing yourself to any expenditure you should think seriously about getting your neighbours to put something in writing. Write out a simple note of what you have agreed and ask everyone to sign it, for example:-

We agree to have J. Smith & Co. repair the chimney head in accordance with his estimate of 12 May 1992 for £106.50 including. VAT. We the undersigned agree to pay our share of the costs.

This generally has the effect of making everyone realise what they are going to have to pay. If you are embarking on a big job you would do well to draw up **minutes of agreement.** We have printed two examples of a minute of agreement and you can change them to suit your situation *(see appendix 4).* The **first minute of agreement** should be used when you are agreed on carrying out repairs and may have to incur some costs before you can take the repairs any further. This can be used to show builders and architects that you have the backing of all owners in getting repairs arranged.

The second minute of agreement should be used when you have quotations from builders or an architect's specification but have not yet committed yourselves to a builder. These letters are not designed to cover every legal loophole. They are designed to provide written evidence of someone's intentions and it would be difficult for them to try to wriggle out of paying after having signed these minutes.

Paying For Repairs

It is easy for disagreements between neighbours to occur and you should be guided by your Title Deeds. You may want to make a once only change to cope with repairs if the deeds seem unfair. Changing the title deeds will usually require a solicitor (see chapter 2), alternatively you can make changes just for the repair work provided you get everyone's written agreement. You may wish to use one of our **model minutes of agreement** (see appendix 6).

When organising a common repair scheme, watch out for the following problems:-

TV Aerials: if you are putting up a communal mast or aerial, will everyone pay a share or only those who have TVs?

Windows: if you do not ask the

contractor to give individual costs for windows, you will all end up paying the same regardless of the original condition.

Internal Decoration: repair problems and repair work may affect the inside of some owners' flats. Will you include the cost of plaster work and/or redecoration in the common cost or will you leave it to individuals to sort out? Will the grant cover decoration?

Check that the builder will give you separate costs for individual items when required.

Important

You should never commit yourselves to a builder until:

Everyone has confirmed they will pay OR

The council has given you written grant approval.

Paying For Big Repairs

Before starting a big repair you should ask your local Housing Department what grants are available to help with the cost of the repair. You should also check when any grants will be paid. If you need to pay the contractor before you get your grants, you may need a bridging loan. You should also consider getting

everyone in the tenement to pay their share into a **joint account** in advance.

You will also have to consider how you are going to pay for your share of the repair costs after any grant has been paid. This is a question which may worry many of your neighbours and it is worth asking about loans while you are talking to the council about grants *(see Chapter 8).*

Professional Help

This chapter tells you about employing people to help you get repairs done. Major repairs and building work are usually carried out by a team of people, led by an architect, surveyor or perhaps your factor.

Do you Need Professional Help?

The main disadvantage of using a professional is that you will have to pay a fee. This will probably be in the order of 10-15% of the total cost of the repairs. A professional may also recommend that more work is required than you had intended. Before coming to a decision, you will need to consider:-

The repairs needed: Do a survey. If you need lots of repairs, or a large repair costing more than £2000, or if the repairs are major or structural, then you are advised to get professional help.

Grant conditions: If you are aiming to get a grant, your local council may insist that you have professional help. Ring them and check.

Your capabilities as a group of owners: Some architects are very willing to advise owners on getting themselves organised and getting grants. This would be very useful if you are total novices as far as repairs are concerned.

Architects

Architects will normally have completed a University or College course of 5 years study and 2 years supervised working in an architects office. An architect who has completed a degree course should have the letters B.Arch. (Bachelor of Architecture) following their name. However, not all architects use these letters, even though they have them. In order to use the title 'Architect' in the course of business, the architect's name must be on the register of the **Architects Registration Council, UK** (ARCUK). Many architects also have the letters RIBA (or ARIBA or FRIBA) after their names - this indicates that the architect is a member of the **Royal Institute of British Architects**. Scottish architects have the letters ARIAS or FRIAS after their names. This stands for Associate or Fellow of the **Royal Incorporation of Architects in Scotland**. Members of RIBA or RIAS may call themselves **Chartered Architects**.

Someone calling him or herself an 'Architectural Designer' will not have the full training that an Architect or Chartered Architect has (see also Chapter 9 for problems with architects).

There is a professional code of conduct which your architect will follow. You can expect an architect to give you independent advice and take a professional approach to a job, putting thought and care into the specification that will go to the builder. You can expect the architect to try to get value for money. *For example, if your roof repair requires scaffolding to be put up the architect may suggest that stonecleaning and repointing can be done at the same time as the repair at a cost which is less than the cost of the two jobs done separately.*

Surveys

You can expect the architect to carry out a survey of your building but you cannot expect that all hidden defects will be found (this would entail extensive opening up of the floors, walls and hidden areas). However, an architect may suspect some hidden problems and will allow for a **contingency sum** to cover the cost of such unforeseen items that may arise.

Choosing contractors

For jobs being planned in plenty of time, you should expect the architect to get estimates or tenders from a number of contractors. You can expect the architect to advise on the selection of contractors who will be most suitable for the job in question. They may have some knowledge of the contractors' working methods and performance but you cannot expect the architect to come up with a perfect builder working at knock down rates. On some occasions it may be appropriate to go directly to one builder, i.e. where a job needs to be done quickly **or** is a very small job **or** is one which is very specialist.

Inspecting builder's work

You can expect the architect to make regular visits while work is in progress to check that work is being done according to the specification. You cannot expect these visits to be more than once a week or once a fortnight. You cannot expect the architect to give detailed supervision of tradesmen and quality of workmanship. This is the builder's job. However, you can expect the architect to put your complaints to the builder at regular intervals. More detailed inspection is often done by a clerk of works.

Dealing with owners

You cannot expect an architect to apply for grants for you, make arrangements for rehousing you while work is going on, or move furniture. However, many architects will be able to advise you on how best to solve such problems.

Building bureaucracy

In all major building projects, there are a lot of regulations, approvals and procedures to go through before work starts on site. Your architect will advise you on the steps to be taken and will usually make the various applications on your behalf (see chapter 5).

Whose side is the architect on?

When the architect is preparing a survey, or specification you will be the only client. When you go to contract, however, the position changes and the architect will become a neutral party, administering the contract fairly between yourself and the builder. Under the terms of a normal building contract only the architect can issue instructions to the contractor. Your architect will report to you on matters of progress, and any unforeseen circumstances on site. Payment to the contractor is made after the architect issues a certificate, and the client is bound by the contract to meet the payment, within a predefined period.

What will they charge?

An architect will charge between £30-£45 per hour plus expenses for jobs which have not gone to contract (senior architects or partners may charge £50-£100 per hour). For jobs going to contract you will need to negotiate an appropriate fee with your architect, usually on the basis of a percentage of the construction cost. The RIAS do publish in 'Architect's Appointment' **recommended fee scales** which you may wish to consult before accepting any fee quotation. This is now being superseded by The Standard Form of Agreement for the Appointment of an Architect (SFA/92). The SFA guide publishes a Historical survey of fees charged for Basic Services Commission between 1985 – 1990. You will pay extra for a quantity surveyor, structural engineer and clerk of works.

The percentage rate falls as the cost of the job goes up, so you may find it worth getting together with other tenements in your street to get a job done more economically.

How to Find an Architect

The RIAS produce *'a guide to Chartered Architects and their services'* which lists all the architects in Scotland and gives the type of work usually handled by the firm. The RIAS will advise you on choosing a suitable architect and can give you a short list of appropriate firms.

Professional Bodies

Royal Institute of British Architects, 66 Portland Place, London W1N 4AD (RIBA)

Royal Incorporation of Architects in Scotland, 15 Rutland Square, Edinburgh EH1 2BE. Tel: 031 229 7545. (ARIAS, FRIAS)

Architects Registration Council of the United Kingdom, 73 Hallam Street, London W1N 6EE. Tel: 071 580 5861. (ARCUK)

The following publications are available from the RIAS bookshops at 15 Rutland Square, Edinburgh EH1 2BE. Tel:031-229 7205 and 545 Sauchiehall Street, Glasgow G3 7PQ. Tel:041-221 6496.

Architects Appointment - describes the services, fees and charges. Price £3.00 plus postage.

You Wish To Build? - a general leaflet giving a simple outline of the course of a normal building commission. Free.

Building Surveyors

Much of what we have said about architects also applies to building surveyors. There are a number of branches of the profession which carry out other tasks such as:

Valuation surveys.
Structural and defects survey.
Property management.
Reports on repair work.
Building valuations for insurance.

A building surveyor will make similar charges to an architect when carrying out repair work.

Surveyors will have taken exams set by the **Royal Institution of Chartered Surveyors** (RICS), either by working in an office and taking part-time classes or by completing a full- time course at a University or College. After taking the exams, surveyors will have to do two years work in an office and a professional practice examination, before being able to use the letters 'ARICS' (Associate of the Royal Institution of Chartered Surveyors) after their name. Surveyors who have been in the profession a lot longer will have the letters FRICS (Fellow of the Royal Institution of Chartered Surveyors) after their names. About half of all surveyors are members of the RICS.

The RICS have a disciplinary procedure which aims to prevent unprofessional behaviour (for example, accepting bribes), rather than incompetence. They will also offer to find arbitrators.

You can find surveyors by looking in the Yellow Pages under **Surveyors - Building** or by asking the RICS for a list of members in your area with appropriate experience.

Professional Body - *Royal Institution of Chartered Surveyors (Scottish Branch), 7 Manor Place, Edinburgh EH3 7DN Tel:031-225 7078.*

Clerk of Works

Clerks of works carry out the detailed inspection of building, the assessment of repair needs and the assessment of builders' work.

You can expect a clerk of works to spend a specified time on site, preferably allowing enough time for a daily visit. Don't expect to know when the clerk of works is coming as the surprise visit is often the best way of detecting the builder doing something wrong. Sometimes, however, the clerk of works will arrange appointments with the builder to inspect work before it is covered up.

The clerk of works may be employed and supervised by the architect or surveyor so check with them what arrangements will be made on your job. Some clerks of works are self-employed. An architect may be able to suggest some names. You can also contact the **Institute of Clerks of Works** in Great Britain and ask them to suggest some names.

Most clerks of works should carry their own indemnity insurance cover. If they have passed the Institute exams they can become a member of the Institute of Clerk of Works and use the letters ICW after their name.

If you have problems, and the clerk of works is employed through an architect, you should take up any complaints with the architects. If you have employed a clerk of works yourself, you will need to consider whether to take action on grounds of **negligence** or **breach of contract.**

A Clerk of works will cost, on a contract, 1%-1.5% of total contract value and on an individual job basis,£20-£30 per hour.

Professional Body - Institute of Clerk of Works of Great Britain, 41 The Mall, London W5, Tel:071 579 2917.

Quantity Surveyors

Quantity Surveyors estimate quantities of materials needed for building work (and therefore can help judge how much a job should cost). They also deal with contracts and tenders and drawing up Bills of Quantities. The Bill of Quantities is a list giving details of the numbers and sizes and quantities of every part of the job. The contractor prices against this list and it is constantly referred to during the building work.

The cost of using a quantity surveyor depends on the extent to which they are involved in the work, and can vary from 3% to 6% of the building cost. The quantity surveyor will work directly with the architect, drawing up the Bill of Quantities, attending site meetings and checking how much work has been done by the builder. The quantity surveyor may also determine the payments to be made to the builder during the course of the work.

Professional Body - *Royal Institution of Chartered Surveyors (Scottish Branch), 7 Manor Place, Edinburgh, EH3 7DN. Tel:031 225 7078.*

Structural Engineers

Structural engineers give specialist advice on structural and foundation problems. They are normally called in by the architect if a particular structural problem is found, e.g., problems with the foundations, bulges in the walls.

The structural engineer will visit the property and examine the problem carefully. Sometimes the engineer will need to decide if the problem is an old one for which no action is required or one which will continue and must be rectified. Often they will monitor a crack over a period of time to see if there is further movement. If there is danger, you will be moved out immediately. You will find that the structural engineer costs between £30 and £45 per hour (senior engineers or partners may charge £50-£100 per hour).

Professional Body - *Institute of Structural Engineers, 11 Upper Belgrave Street, London SW1X 8BH. Tel: 071 235 4535*

Choosing the Right Person

Make up a list of possible firms from:-

Professional organisations (listed above)

References from other people who have had work done

Yellow Pages

Ring the firms concerned and check:-

Whether they are used to the type of work you want

What fees they charge

Whether they will help with applications for grants and planning permission.

Whether they will tell you the names of similar jobs they have done so that you can check them out.

When you have narrowed the list down to two or three firms, arrange a meeting with the architect, and check whether:

You will get on with the architect.

The firm will attend evening meetings.

The firm will go ahead even if you still need to get one or two owners to agree.

The firm will first carry out a survey and give estimated costs before you commit yourself to the work or using that firm.

Other professionals will be needed and what their fees will be.

A particular individual in the firm will do your work, and what qualifications the person has.

You can have regular meetings with the firm when the builder is working on the job.

You will be charged for copies of letters plans or other expenses.

Working With Professionals

Before you employ an architect or surveyor make sure you both understand whether your top limit on costs includes fees, VAT and the cost of applying for building warrants, etc. These costs can add 30% or more to the basic building cost.

Make sure you understand that once the building contract is signed the professional's job is to be fair to both sides when administering the contract. The professional is not acting only on your behalf.

Lack of communication is a general problem and leads to many other misunderstandings with the professional. Try to arrange regular meetings.

The following tips may help you get on better with your chosen professional:

Clarify the method of communication between you and the professional and agree the frequency of reports and meetings.

Clarify the procedures for decision making, particularly about expenditure.

Make sure you understand who is responsible for subcontractors (*see What Standard Building Contracts Say in Chapter 5*).

Don't be forced into making instant decisions. Make certain the professional knows you need more advice and more time to think things over.

Always confirm things in writing; professionals are normally under an obligation to write to you about changes, monitor this.

Check whether any changes will cost you extra money before you agree to them. If you cannot avoid changes which cost extra, ask if the professional can propose savings elsewhere, and advise you of the consequences of implementing them.

Don't presume that all work proposed will be grant aided and ensure at all times that the grants department is kept aware of the work. Make sure grants are properly explained to you.

Organising a Repair Scheme

What the Architect Does.

1. The architect will carry out a survey. From this survey, the architect should be able to give you an **indicative** cost. This is the architect's estimate of what you will be charged by the contractor. It is not the same as an estimate given by a tradesman for a job. Check whether the indicative cost includes fees and VAT. The architect will require all owners to agree to this **indicative cost** before he proceeds any further.

2. If you need **planning permission** the architect will prepare the application and apply for it in your name, at the appropriate stage. Be ready to pay your share of the cost of the application.

3. The architect will now carry out a detailed condition survey and start to draw up the detailed **specification**. This is a description of what the contractor is expected to do. Some architects use standard specifications which are the outcome of their experience dealing with repair jobs.

4. The architect will apply for a **building warrant** (about building standards, fire resistance etc) at this point if it is required. Again, be ready to pay your share of the cost of this application.

5. If the project is a major repair, the drawings and specifications may be sent to a quantity surveyor. The quantity surveyor's job is to work out the quantities of materials needed to complete the job. This is known as the **Bill of Quantities.** Some jobs will not need a Bill of Quantities and the contractor will be expected to estimate quantities himself.

6. The architect will advise you about the preparation of a list of contractors who will be asked to **tender.** The architect should get the clients' approval for the list as you may wish to delete the names of firms you particularly do not want to use. At this stage architects will ask their clients' permission to go out to tender. This gives clients a chance to change their minds without incurring additional costs.

7. The **plans, specifications** and **bills** will now be sent to contractors on the list for tendering. Each contractor will work out what they will charge for doing the work, on the basis of the drawings. This is their tender. Contractors are given about 3-4 weeks to tender. A date and time is set for the return of tenders. No tenders will be opened until this time.

You may ask to be present when the tenders are opened.

A formal procedure for opening tenders is adopted in order to ensure fairness.

8. Once the lowest tenders have been checked for accuracy by the quantity surveyor, the contractor can be chosen and the contract awarded. You may need to get local authority approval to accept the contract if you are getting grants etc.

The contract is very important. *We explain the terms of normal building contracts in Chapter 5.* Before the work goes on site some architects take photographs of the interior of the building so that they can assess damage allegedly caused during the contract in comparison to damage caused prior to the contract.

9. The architect should hold a meeting with the owners at this stage and advise you of expected building conditions, to check insurances, etc. There will also be a **pre-start meeting** with contractor to check firm's insurance, start dates, and other details, for example, access to the building.

10. When on site, the architect will hold regular site meetings with the contractor to assess progress and resolve any problems. You may ask to receive copies of the minutes of these site meetings. In grant-aided repair work, the District Council will also inspect the works, although usually not till near the end of the contract. At regular intervals throughout the contract, the architect will certify payments to the builder by issuing interim certificates. The builder must be paid the sum on the certificate otherwise the client (you) will be in breach of contract. Once the work is completed a practical completion certificate can be issued and half of the retention sum is released to the builder.

11. After practical completion, the builder is still responsible for correcting defects arising within the Defects Liability Period. This can last for up to a year after practical completion. The architect should inspect the works at the end of this period and ensure the builder corrects the outstanding defects. The remaining part of the retention sum is released when all the defects are made good.

Work starts on site.

Dealing With Builders

The Building Contract

If you have employed an architect to act for you, your builder will probably be employed under a standard building contract. In order to ensure fairness to both builder and client the contracts are based on plans, schedules, specifications and bills of quantities drawn up by your architect. These documents contain all the information the builder requires in order to price the job accurately, and are usually sent to several builders. This is known as tendering. The contract is normally awarded to the builder giving the lowest price.

There are reasons for not choosing the lowest priced tender, e.g. if the builder has already accepted other jobs they may be too busy to carry out your job to the standard and time scale you require. Your architect should advise you when they suspect this to be the case.

You or your architect and the builder will sign the contract or there will be an exchange of letters between the architect and the builder. Most people never actually see the contract which is being used. The contract puts obligations on both you and the builder. Once the contract is made, you cannot make big changes or change your mind about going ahead with the work. If you do, you will probably have to pay compensation to the builder. If a Section 108 Notice or Agency Agreement is being used, the contract may be between the District Council and the builder.

Choosing a Builder

If you have not employed a professional to manage the repair work, you will have to employ a builder yourself. Before choosing a builder, you will need to draw up a specification of the repairs to be carried out (the information from chapter 13 onwards will help with this).

In choosing a builder, you should never forget that you are buying a repair. You may want to support your local trader but most of the time you will want to shop around to get the best buy. The most important consideration should always be for reasonable quality and reliability rather than the cheapest price.

The building trade is notorious for the number of "cowboys" among its members. Tradesmen tend to come and go especially in the large and medium sized firms. Many tradesmen are self employed and work under contract for several firms. It is difficult to be sure that you are going to get quality or reliability but you can take some steps towards achieving value for money:

First, you should compile a list of possible firms.

Secondly, you should ring these builders and check whether they will actually do your job at the time and under the conditions which suit you.

Thirdly, you should check a bit further on the firms, for example, ask a builder to show you examples of previous work.

Finally, you should ask three or so firms to give you a detailed written quotation.

The box overleaf gives you an idea of some of the questions to ask.

Getting a Rough Cost

A rough cost will give all owners an idea of how much money they will need to raise and will help you decide which items on your list are top priority and which are a lower priority. There are several ways in which you can get rough estimates:

1. If you are fairly certain that you will get repairs off the ground, and you are prepared to pay, you could go to an architect or surveyor and ask for a **preliminary survey** and **indicative cost.** This will also tell you what needs done as well as what it will cost. This is probably the most accurate kind of initial estimate that you can get but it could cost you. This need not bind you to using that or any other architect or surveyor.

2. Although some contractors may be willing to give you a rough estimate over the phone, don't place too great reliance upon its accuracy. Such estimates are not recommended - it is better to invite at least one contractor to visit the

building to assess the work, however rough the estimate, than to rely on a phone estimate.

3. You could approach people who have had similar work done recently. They should be able to give you the cost they have paid. Remember to ask for the total cost of the work, rather than just their flat share. You should also allow for tenements being of different sizes.

4. Your local District Council may be able to carry out a survey of your house (especially if you are looking for a grant) and they may also be willing to give you a rough estimate of costs. (The council may charge for the survey - ring first and check).

You may be very surprised at the high cost of repairs. It is estimated (June 1992) that it currently costs a builder £500 a week, i.e. £100/day to employ a craftsman and this figure does not take into account the cost of materials or profits to the firm.

Finding The Good Builders in Your Area

Stage 1:
Finding Your Builder

Word of Mouth

This is undoubtedly the best way of finding a reliable building firm. Neighbours and friends may have used a good tradesman and be willing to recommend them. Don't listen to people who refer you to their 'brother-in-law' unless you've seen him do good work. This is one of the more popular ways of losing friends.

Yellow Pages

Someone who is in the Yellow Pages is unlikely to be a fly-by-night. You will still need to be careful though - ask for references, check the firm out, and ask for a list of prevoious projects you can go and visit.

Newspapers and Newsagents

Newspapers and Newsagents' windows are a good source of local builders. You will have trouble recognising the reputable builder from the "cowboy".

Trade Associations

Most trade associations expect a fairly high quality from member firms and any firm displaying a trade association mark on its notepaper is likely to be reputable. It might be more expensive but get estimates and check. Trade Associations will normally take up complaints on your behalf. The trade associations you may come across are:

National Inspection Council for Electrical Installation Contracting (NICEIC), Alembic House, 93 Albert Embankment, London SE1 7TB Tel: 071 582 7746.

Electrical Contractors Association of Scotland,(ECAS) Bush House, Bush Estates, Penicuik. Tel: 031 445 5577.

Scottish Building Employers Federation,(SBEF) 13 Woodside Crescent, Glasgow G3 Tel: 041 332 7144.

Scottish Master Wrights & Builders Association, 26 West Nile Street, Glasgow G2 Tel: 041 221 0011.

Federation of Master Builders, 33 John Street, London WC1N 2BB Tel: 071 242 7583.

British Chemical Dampcourse Association, Building No 6, The Office Village, 4 Romford Road, London E1S 4EA Tel: 081 519 2588.

National Federation of Roofing Contractors, Scottish Regional Association, 13 Woodside Crescent, Glasgow. Tel: 041 332 7144.

Scottish Master Slaters & Rooftilers Association, 13 Woodside Crescent, Glasgow. Tel: 041 332 7144.

Scottish and N. Ireland Plumbing Federation, (SNIPEF) 2 Walker Street, Edinburgh EH3 7CB Tel: 031 225 2255.

Scottish Glass Association, 13 Woodside Crescent Glasgow G3 Tel: 041 332 7144.

Confederation for the Registration of Gas Installers, 4 Marine Drive, Edinburgh. Tel: 031 552 6960.

Stage 2:
Drawing Up a Shortlist

Are they in the telephone directory?

If they are, they will have been in business for at least a year. It will also show if they are trading as a business rather than as an individual.

What is the business address of the firm? If they don't have a business address you won't be able to find them when things go wrong. Check the business address. Go and visit.

If it's a house, you certainly should not think about using that builder for any big job. Where is the builder's yard? If it is nearby you will have to pay less for travelling time between the yard and your project. It has been estimated that up to 40% of costs arise from travelling.

Does the builder handle grant work? Are they prepared to wait to be paid until you get the grant cheque from the council?

Are they prepared to come out and give a quotation? Will they charge you for this? How long will they hold the quotation open for? Is this long enough for you to get the grant approved?

Are they used to doing the kind of work you want? Are they a specialist?

Check the builder's references. Ask the referees if the firm turned up on time, was the firm efficient and friendly? Were there many problems and were they willing to put any defects right quickly? How far out was the final cost from the original estimate? Has there been time for most problems to show up?

Stage 3:
Making the Final Selection

Make sure the firm can show you evidence that they have current **public liability insurance.**

Does the builder know that you are approaching two or three firms for quotations?

How will the work be supervised?

What other jobs are they doing? When would they be able to fit your job in?

What safety precautions will the builder take? *(See Going on Site on page 23)*

How long is the quote/estimate open for?

Has the firm done this kind of work before? Get some references.

How will the builder go about doing the job?

What disruption will there be? Who are you talking to? If a salesman, what do they know about the technical side? Are they paid on a commission basis? Is the salesman permitted to amend contracts to meet your terms?

Will the firm agree to your terms (see the Do It Yourself Contract later in this chapter). Go through these carefully and in detail. Check the 'Going on site' section.

If the Job is Smaller:

Has the tradesman been trained? - If so can you see a certificate for the training?

Check how long they have been in business?

Ask for references of other jobs they have done and check them out.

Have they proper receipt books or notepaper?

Is it VAT registered? (check the notepaper).

What safety precautions will they take?

Has the firm got **public liability insurance** (ask to see a copy of the policy).

POINTS TO NOTE - *Did the builder appear to be efficient, knowledgeable and co-operative? Do you think you can trust the firm? You are very unlikely to find someone who can give the right answer to all these questions but you should at least be able to spot the real cowboys.*

Quotations and Estimates

It is important to get written estimates and quotations and to check the details carefully. Don't be afraid to ask the builders to clear up any queries that you have.

Quotations and estimates are often confused by owners and builders. Strictly speaking, an estimate will give you an idea of costs but the final cost may be quite different from the original estimate. A quotation is what you will be charged for the work and it should not be exceeded. You should make sure that what you get is a quotation. A quote should legally allow for everything that will be required to complete the job. However, additional problems are often discovered after work has started, and you should always ask for an itemised breakdown of costs. This should give you a rate for each part of the job - i.e. for scaffolding; for each square metre of repointing or each linear metre of gutter. For small jobs it may be possible to get an estimate or quote based on time and materials. You may find yourself paying for a slow workman. Ask for an estimate of how long it is expected the job will take.

One other advantage of getting an itemised breakdown is that you will know exactly what is or isn't included in the price. This will make it easier to compare quotes and to adjust the price for any unforeseen extras.

If the builder thinks that what you want to do is not appropriate he may suggest change to the specification. You will find it difficult to compare estimates so ask the builder to estimate for both your own and their specification.

What Causes Differences Between Quotations

When you receive your quotations, check them thoroughly. If you have been sent a quotation for the wrong thing, go back to the builder and ask them to reprice. Check the rate you are being charged - that is, how much it costs for every metre of gutter or to erect scaffolding.

VAT

Since June 1984, all building works (except some to Historic Buildings) are subject to VAT, currently 17.5% (June 1992).

The Profit Element

All builders allow a percentage for profit. If the builder is taking a lower profit margin it may be because they cannot get enough work to get a higher profit but equally it may be because the firm is more efficient or has low overheads.

Low Overheads

Overheads are costs like the firm's rent and rates, equipment and office staff wages. Low overheads may mean that the builder does not have all the resources necessary to do your job. Smaller firms tend to have lower overheads but fewer resources. Some firms may have low overheads but will include more expensive rates for work to offset this. This of course adds to the difficulty of comparing the quotations.

Waiting for an accident to happen.

Estimating Skill

Your builder may not be a very good estimator though you might still get a good job. If the job has been poorly estimated you may find your builder coming back during the job asking for more money. If you've got a good contract you will be able to refuse to pay but the builder may be tempted to cut corners. It may well be better to pay the extra, as long as it is clearly defined and seems reasonable.

Materials

One builder may be using a better quality material than another. You can check easily by ringing a builders' merchant and asking about the cost and quality of the materials named. For more specialist items you should check whether the material meets British Standards or has an Agrement Board Certificate. Again, you can check this with a builders' merchant.

Health and Safety

One builder may be trying to avoid using more expensive but safer working methods. If you've got itemised quotations, you will be able to see what you are being charged for scaffolding or for using two men rather than one on awkward jobs.

Safe working methods can make a job more expensive but builders can be thrown off the job by the Health and Safety Executive for failing to work properly and safely. This ends up costing

owners more. The Building Control section of the District Council may be able to advise you whether a builder is attempting to cut corners over safety.

You may ask who is going to report your builder to the Health & Safety Executive if they are cutting corners with safety. The answer is any other building contractor driving past. A reputable firm loses jobs to those who avoid taking safety precautions and offer lower quotations. The reputable firm will get more business if they can stop these unfair practices.

If you suspect that the builder is not taking due care and attention, then do not hesitate to contact the Health and Safety Executive.

Your Contract with the Builder

The Common Law

Whether you have signed anything or not, you have a contract when you ask a tradesman to carry out work and they have agreed. Even with a verbal agreement, both you and the tradesman are bound by the **common law of Scotland**. This law states that you can expect to get:-

Work of a reasonable quality carried out in a tradesman-like manner.

Materials of a quality appropriate to the job finished in reasonable time.

In return the builder can expect to be:-

Given access to the work at the necessary time.

Given instructions in reasonable time.

Not to be obstructed in their work.

Given reasonable payment for the work done.

The common law gives you some legal protection but a lot depends on what 'reasonable' means. A written contract is the best basis for a good working

Getting Permission

Planning Permission

Strictly speaking you need planning permission to alter a building or land in any way. The simplest thing to do is to ring your local District Council Planning Department, tell them that you live in a tenement and ask whether you will need planning permission for the work you are carrying out. The planning department may want to impose special regulations on styles of windows if you are replacing them. You should also check on stone cleaning or stone painting, building walls or bin stores, garden sheds or garages, changing the type of tiles or slates used on the roof and putting in new front or back doors. If you do need planning permission there are scale fees depending on the type of work. You can expect the application to take two or three months to process, and longer if the council is not meeting over the summer months.

Building Warrants

You will need to apply for building warrants for such works as:-

Knocking down walls or other structural alterations.
Installing new bathrooms.

Major repair works

Again you should check with the local building control or technical services department of the District Council. You will probably need to supply properly drawn plans and will have to pay a fee based on the estimated cost of the work. The building control officer may be able to give you some useful suggestions about your work. They should also check the work at the end of the job and you may find this check will pick up defects which you yourself did not notice. Indeed they may sometimes require work to be done which was not originally priced for!

Electricity

If you are installing new fittings which will increase the amount of electricity which you use, for example, installing an electric shower or rewiring, you will need to tell the local Electricity Board. They will then check to see if the existing supply to your house is adequate. If it is not, they will provide a new supply, normally free of charge. Ring the Electricity Board or call at one of their showrooms for an *'application for supply'* card.

If you are rewiring, you will need to get the Electricity Board to check the wiring before it is connected to the Electricity Board's fuse box. If the wiring is dangerous, the Electricity Board will not connect up and they will charge for a second visit. Ask for a *'test of wiring'* card. Beware of partial rewirings which try to avoid new connections to the fuse box - they will probably overload the old wiring - to your danger.

To avoid cowboys, use a firm registered with the National Inspection Council for Electrical Installation Contracting (NICEIC).

Gas

You will need to apply to get British Gas to alter any gas pipes leading up to the meter. You may call on your own tradesman to do any other work but the **Gas Safety Regulations** say that any work must be done by a **Corgi registered tradesman.** You cannot do the work yourself. The Gas Board will call and check appliances if you suspect a gas leak or any other danger, for example, inadequate ventilation in the room for the gas appliance. The Gas Board will provide up to 30 minutes work without making a charge and this is usually sufficient time to make any appliance safe.

If you are an elderly or disabled person living on your own you are entitled to a free safety check.

Water

If you are replacing lead pipes with a grant you will need to check with the Water Authority that you are using the right materials. You will also need Water Authority permission for new installations. This is a basic check to see if appliances meet the bye-laws. The Water Authority are likely to check on work afterwards to see that there is no waste or contamination of water - they may raise issues about access to tanks etc, but do not usually check on the quality of work.

Skips

If you are using a skip you will need permission from Building Control and you will also need to make sure the skip is properly lit at night if it is on the street.

Scaffold

If any part of the scaffold is on the street, that is, not in your front garden - you will need permission from Building Control. You may also need to light the scaffold; protect pedestrians from bumping into the poles; erect fan scaffolding to protect people from falling debris.

relationship with your builder. It lets both sides know exactly where they stand. The common law is no substitute for a detailed written contract. If you know a really good builder who has done work for you before, then the common law may be sufficient.

Using Your Own Contract

If you are employing your own builder, you can use some of the clauses in our **'Do it Yourself Contract'***(see appendix 6)*. Discuss the contract with the builder first and make it quite clear that your aim is to resolve any problems which might arise in the course of the work rather than to handicap the builder with red tape. You can put the appropriate clauses in a letter of instruction and ask the builder to confirm in writing that they agree to keep to these instructions.

Using a Contract Drawn up by the Builder

Many specialist firms have a **standard contract** which you are asked to sign. Read the contract carefully before signing it. Don't rely on the salesperson changing the contract for you. Many contracts have a clause which states that they may only be changed by a director of the firm. Get a letter confirming the changed contract.

Beware of contracts which include the statement, *'subject to grants being available'*. Many firms selling replacement windows advertise this as a feature of their contract. The phrase usually means *'subject to grants being in existence'*, it does NOT mean *'subject to you personally obtaining a grant towards the cost of having your windows replaced'*. You may get some protection from the **Unfair Contract Terms Act** or the **Trades Description Act.** However, the protection is fairly limited. You cannot rely on the law to protect you. Read any contract carefully and do not sign anything you are not clear about. If in doubt, get advice.

Building Bureaucracy

To carry out alterations to buildings you need to get official permission. Some people think this is needless interference but everyone is put at risk when a whole tenement is made unsafe *(for example, because someone has knocked down a structural wall without proper permission and normal precautions).*

Check whether you need any of the permissions described in the adjacent box. If you have asked your builder to get these permissions, make sure they have. You will be responsible as the owner if they have not. The local authority will take action against you in the first place and you will then need to take action against the builder.

Use the Bureaucrats

If you have to obtain permission for carrying out work, you will probably need to have the work checked to see that it complies with the regulations. This inspection will at least tell you that the work is not dangerous and you can use it as a very rough check on the tradesman. You should always be there when the inspector calls - you may get some other free advice. *You should not rely on this as a final check, however. For instance the Electricity Board will check wiring before it is connected to the mains street supply BUT they won't tell you whether the wiring has been properly plastered over.*

Going On Site

Are you ready for the builder? Make sure you know who the foreman, boss or site agent is and what their emergency phone number is. It is a good idea to ensure your house and possessions are protected from damage or theft during the works.

Check the following:-

1. Does your insurance cover you? Write and ask. If you have a mortgage, you may need to go through your building society unless you have organised your own insurance. Check both buildings and contents policy. If mortgage related insurance cannot be extended you will have to ask the builder to take it out for you as part of the contract.

2. If you don't have **locks on your windows**, and scaffolding will be outside, it may be an idea to nail up your windows temporarily.

3. Your garden could get badly mangled. Move or protect your plants.

4. If workmen are coming in the house, **put everything valuable or breakable in a safe place**. Will you need to move furniture? Have you got dust sheets for furniture and carpets or will the builder be providing these? You could club together with your neighbours to buy a

large roll of plastic sheeting. Check under 'plastics' in the Yellow Pages or go to a builders merchant.

5. Decide between you who will be the **close/stair representative** for the builder to speak to about access, additional costs, changes in specification etc. Make sure all owners know that only the chosen representative should give instructions to the builder. This will save the builder being given conflicting instructions.

If you have employed an architect, they will be responsible for giving any instructions to the builder. If you have problems tell the architect immediately. You may request regular progress reports on the work, or agree times to meet with you to discuss any problems. Always try to ensure everyone is kept informed and that everyone understands the method of communication.

6. Moving. What will you do if anyone needs to move temporarily *(see Rehousing on page 54)*. If people cannot move speedily, this may hold up going on site. Make sure the builder is kept informed of when people are moving. Builders will not automatically know that someone has moved and that they can therefore start work.

7. Check your ceilings. Old tenements have old ceilings and these can come down if there is a lot of hammering. You can't do repair work without hammering so check the ceilings by prodding them with a broom stick. If the plaster moves or if there are cracks, tell your architect or builder. You may need to support the ceiling while work is going on. Whatever the condition of your rooms and ceilings, take a photograph before work starts - it could be useful evidence later if damage occurs during the building work. Normally your architect would do this with the builder at the start of the contract.

8. Gas fires and boilers. If work is being done to chimneys you will not be able to use gas fires because the flues may get blocked. Make sure you have some alternative form of heating and get the chimneys swept before using the gas fires again.

9. Disruption. You will not be able to lead a normal home life while building work is going on. Migraine sufferers, night shift workers or families with young children should try to make arrangements to stay elsewhere during the day. The Social Work Department may be able to help with temporary day time accommodation for the elderly or disabled.

10. Toxic Fumes. If you are having timber treated you will all need to stay away for a few days, as some chemical treatments require at least 48 hours, evacuation. This applies not only to the house being treated, but adjoining houses as well. Ask your builder about dry and wet rot treatments, as well as woodworm treatment. Do not believe them if they say the chemicals are harmless, ask to see the manufacturer's leaflets *(Lindane is a toxic chemical used in woodworm treatment, although it has been banned in Sweden it is still used in the U.K.).* Warning notices with health and safety precautions needed should be fixed to the containers by law. If you are in any doubt ask your architect for information on the chemicals.

11. Dust Protection. Dust from old plaster ceilings, general building work and stone cleaning can cause havoc with sensitive electrical goods. Try to remove such items from the affected rooms before work starts. Most cleaning of sandstone is now done by the wet grit method or by chemical cleaning, but even so, dust can be a problem. The builder should sheet and tape the joints on all windows, and external doors from the outside. Your roll of plastic may come in useful if the building is being stone cleaned. Hang the plastic over all your windows and stick it down well all round the edges with masking tape. This will stop the sand and dust from coming in through the windows. Ask the builder if he will do this for you.

12.Protection from Weather. If slates are being stripped from the roof, the builder should cover the roof with a tarpaulin each night to stop rain coming

in. Inevitably, if a wind blows up, the tarpaulin can lift and rain will come in underneath. You should use the builder's emergency phone number in such cases. Top floor flats can expect to get wet but owners should try claiming on their insurance if the builder will not compensate you.

Living on a Building Site

Normal working life for a building contractor is NOT normal for you. You will have to accept the builder's normal working arrangements unless you have agreed otherwise beforehand *(See the Standard Building Contract in appendix 5).* If you try to change these arrangements when the builder is on site you will run the risk of the builder walking off and not coming back.

Safe Working Practices

If you think anything could injure a small child then it is likely to be dangerous. It is estimated that each year as many children die on construction sites as die on farms and railways. Particularly you should look out for:

> *Inadequate boarding of unprotected holes*
>
> *Erection of barriers*
>
> *Removal of ladders (scaffold looks like a climbing frame to kids)*
>
> *Materials should not be stacked so that they can be pulled over or climbed on*
>
> *Unsafe electric cables*

Do not rely on notices which toddlers cannot read and adults ignore. If you are using an architect, they should be looking out for any hazardous fixings. They should check that the builders are using the appropriate scaffolding on roofs and using catwalks inside the roof space. A foot (or even a whole body) coming through your ceiling is not good for decoration apart from the damage it causes to the worker.

If you think anything is dangerous call the **Health & Safety Executive** or your **Building Control Department.**

Further Reading:

Health & Safety Executive Guidance Note GS7 Accidents to Children on Construction Sites. Price £2.00.

Health & Safety Executive Guidance Note HS(G) 33 Safety in Roof Work. Price £1.50 Available from HMSO bookshops.

Access

If the workers cannot get into your flat on time you cannot expect them to do the work on time. You should expect 24 hours notice but you could ask for more. Make sure your close/stair representative has keys or knows how to get in touch with you. It is worth trying to be in when workmen call so you can keep an eye on them and their work. It does happen though that workmen sometimes do not turn up on time. There is not much you can do but you could write and complain to the builder. Also you could refuse to consider subsequent claims made by the builder for delays due to lack of access.

Facilities

Your builder will need access to electricity, water, toilets and telephones. They should provide all of these things themselves but you might come to an arrangement which knocks a few pounds off the bill. Sort this out beforehand. If one owner's electricity is being used, take careful meter readings and get agreement to pay for usage recorded. Never feel obliged to offer these facilities to the builder.

Tools and Machinery

Make sure the builder locks away tools and disables machinery at the end of the day. You don't want children playing with the machinery or house breakers using the tools. Check that cartridges from cartridge operated tools are not left lying about or in the rubbish. You will frequently find one or two unfired cartridges in a 10 shot strip.

Materials

The builder may want to leave materials on site. It is their problem if things get stolen, but you should help by calling the police if you see anything suspicious. Some materials are inflammable. These should not be left in houses overnight and should preferably be locked in a metal box. This also applies to gas canisters.

Scaffold

If a builder is working at a height of over 30 feet, or is working on a roof, scaffolding **should** be used. No builder will use scaffolding unnecessarily. For odd jobs on the roof, the builder may use a harness and should board the bottom edge of the roof to stop things falling off. The bottom stage of the scaffold should

be boarded if it is over a pavement. It should always be boarded over the close entrance. There should also be a platform where men are working to stop them (and bricks) falling. The scaffold should be tied to the building either through a window opening or with rings and bolts. It has been known for the scaffold to peel away from a building.

Sometimes a builder will use a scaffolding tower. In this case, the measurement of the base line of the scaffold should not be less than one third of the height of the tower. The builder may use 'out riggers' on the scaffold tower to increase the height.

The Health & Safety Executive do not allow materials to be thrown down from scaffolds. Only in certain instances will it be accepted. The builder should be asked to provide chutes or carry it downstairs.

Checking on Workmanship

If you have heeded these warnings, you will have arranged for work to be professionally checked throughout the job and at the end. It is wise to check the work at various stages so that mistakes cannot be covered over. You should also ask to be given about 3 days notice of scaffolding being removed so that you

can get your architect, surveyor or clerk of works to check the work from the scaffolding. If the builder has made mistakes in the work, you will find it much more difficult to get them to rectify the defects if the scaffold is no longer there.

You should also keep an eye on the work yourself. This does not mean hovering over the workmen as you are bound to put them off if you do this. Rather you should have an occasional look at different times of day and stages of the contract. **Occupiers should not use scaffolds**. A great many scaffolds do not comply with the law and some are downright dangerous. Try checking from the top floor of the facing house using binoculars. This will usually show most defects.

Awkward Jobs
Some types of work involve a lot of disruption.

Lead pipe replacement and any replumbing requires removal of kitchen units and bathroom fittings and lifting of floorboards. If fittings are removed, they will rarely go back into place as well as they did before. Check before hand what you will do if there is damage. You

should always have essential facilities reconnected at the end of the day.

Dry rot and damp proof courses are messy jobs requiring removal of floorboards and plaster work. The chemicals used are smelly, inflammable and sometimes toxic. Ask to see the container holding the material to see if the builder is doing as he should. If there is anything you are unsure about, check it. It's your health, and anyone with a chest or breathing problem could be placed in considerable distress by the fumes.

Structural work varies from job to job but you should find out beforehand what it will involve.

Check all openings in the floor are closed over before the builders give you the use of any room.

How Long Will it Take?
It is difficult to be precise about how long any repair will take because it depends on the amount of work which has to be done, whether materials are available, whether you find any unforeseen problems, bad weather etc.

A builder will often do more than one thing at a time. This means that a complete repair job consisting of reroofing, stone cleaning, repointing, repairing chimneyheads, replumbing and redoing the whole close can take anything between 5 and 15 weeks. A major common repair scheme could take between 4 and 6 months.

Your builder or architect should be able to give you more accurate details.

Expect delays if you discover rot or if there is bad weather. If you keep changing your mind, you could cause delays too.

When Things Go Wrong

Workmanship is a common source of arguments between architects, builders and their clients. Things go wrong in different ways and different people are responsible. "Bad workmanship" may relate to:-

Defects - or repairs which do not work as intended, e.g. a new window which won't open. This is the builder's responsibility. Professionals should inspect the work generally and report any defects to the builder.

Work not to specification - where the builder has not done what the professional instructed. This is the builder's responsibility and the professional should sort these things out with the builder.

Inadequate specification - this is the architect's responsibility e.g. specifying the wrong kind of plaster for walls treated with a damp proof course injection.

Sloppy workmanship - an effective repair is made but the finish is rough.

Again, this is the builder's responsibility but if the workmanship is below standard the architect can insist the work is redone. It is the interpretation of what this 'standard' of workmanship is which can cause problems.

Not part of the job - for example, a gutter which was not included in the work now starts to leak, or new plasterwork and painting showing up the old. In this case it is up to you to pay for additional work.

Workmens' damage including rain coming in while the roof is off. This may be the builder's responsibility if you can prove negligence. Some professionals will help with these claims, others will not.

Where the builder appears to be at fault, they may be in breach of the contract. Your architect should be able to take action under the terms of the contract and should advise you on the action to be taken.

Problems With Builders

By **common law** you have a right to goods of a merchantable quality, fit for their particular purpose in accordance with the description or sample and with a good title that is honestly purchased. The common law also entitles you to the competent performance of the ordinary skilled workman.

If you have a problem, you may find it difficult to be certain that you have a reasonable case against the builder. For smaller jobs, you could ask another builder to give you a price for putting things right. For big jobs, you could ask a professional to check the repair and write a report on it. This is good evidence if you have to take legal action but you will probably not be able to recover the cost of getting the report done unless you go to court.

Taking Action Yourself

If you do have problems and are not using an architect, the first thing to do is to write, by recorded delivery, to the builder stating your complaints clearly. Ask the builder to meet you to discuss how things will be put right. When you meet, be polite but firm. Blowing your top won't help - it will just cause antagonism. You already have sufficient cause for complaint if work hasn't been completed in reasonable time with reasonable care and skill, using suitable materials. 'Reasonable' is defined as what is normal in the building trade. If your first letter doesn't achieve anything you should think about getting outside help.

Getting Help

If the problem is defects or workmanship, call on the Building Control Department. If the problem is time taken, higher costs than expected or other problems with the contract, see your Trading Standards Department. (see Contact Addresses in Appendix 7) or your local Citizens Advice Bureaux. Tell these people exactly what happened but don't exaggerate the problem, it could only cause confusion when your helper speaks to the builder. Always take the name of the person you are dealing with. Ask what they will be able to do to help you and how long you should wait before taking further action.

Still no luck? Try writing again and state a time limit by which you require action. If you have not yet paid the builder, pay them now but deduct what you think is a reasonable amount for putting things right.

If the builder accepts your cheque for the work minus the cost of putting things right you can probably assume that is the end of the story. In any case, it would probably not be worth the builder taking legal action against you. If you have already paid, or the cost of putting things right is more than the outstanding bill, you may now need to consider taking legal action.

If the firm belongs to a Trade Association they may have an arbitration service, but check that using this service won't prevent you taking other legal action if the decision is against you. It is worth writing to the Trade Association and sending a copy to the builder in any case.

The Trade Association may have a guarantee scheme to ensure your work is completed and the threat of being struck off their books may be enough to get the builder moving. As a last ditch effort, you may consider taking legal action. It is no use taking legal action if the builder cannot be traced, or has no money or assets to sue for.

If you have paid your builder or supplier through a credit card, such as Visa or Access, you may be able to get some refund from them.

The Law

The law does give limited protection to the consumer who has signed a bad contract. The Unfair Contract Terms Act (1977) says that contract terms which try to avoid liability for death or personal injury are legally meaningless.

Other terms which try to exclude liabilities are subject to a 'fair and reasonable' test. The Trades Description Act (1968) says that you should not be given misleading descriptions of the service or the materials to be provided.

The small claims procedure is available now in Scotland. If you are in dispute over a sum of less than £750, you may go to a Small Claims Court. You would not need to be represented by a solicitor in this case.

Supply of Goods & Services Act (1982) is only law in England and Wales, and does not apply in Scotland. Unfortunately a number of large companies use English contracts in Scotland and these contracts are often based on the supply of both goods and services together. In Scotland, these contracts are regarded as two separate contracts - one for supplying the goods (i.e. window units) and one for services (i.e. installing the windows). This makes life complicated, but you should still be protected by Scots common law.

The Repairs Balance Sheet

Once you have an estimate of what your repairs will cost, you will need to consider, not just if you can afford the repair but also if you can afford not to repair. When it comes to paying for repairs, there is a lot more to consider than just the final bill. In this chapter we help you consider all the advantages and disadvantages of repairing.

Costs and Benefits of Repairing

If after considering all the costs and benefits of repairing you decide that you just cannot afford to carry out repairs, you have two courses of action:

Scale down the repairs and only do the most essential repairs (see Preventing Further Damage)

Think about the future of your property. Can you afford demolition? (see Demolition and Compensation). Could you get the District Council or a Housing Association involved to help with costs? (see Housing Action Areas)

some people buying your flat altogether. If you have repair plans underway, or a **Section 108(S24)** Notice has been served on your close, building societies will be happier to lend money.

Much of the value of your house comes from the standing of your neighbourhood. Sometimes, your repairs alone will do little to affect the value of your property. If the whole street or neighbourhood decides to carry out repairs this may have a greater impact on property values.

You could get together with your neighbours to form a residents association to persuade people to carry out repairs and improvements or to put pressure on the District Council to carry out other improvements in the area. People at the community council, your local community centre or your local councillor may be able to give you assistance.

demolition (though you could get a council house). All in all, it is probably worth putting off demolition as long as you can *(see Demolition on page 49).*

If you want to check the length of life of your property you should ask your local Planning Department whether they have any intention of using the land your house stands on for a road, a new school or some other similar project. If they have such plans, you may find that your property is *'blighted'* - it is not worth repairing and it is difficult to sell. Your best course of action may be to form or join a local residents association to seek information and help as a group. The following agencies may be able to help:

Planning Aid is a voluntary organisation which provides advice on planning issues to tenants, Residents Associations and community groups. They can be contacted at:

041 649 7359 (Sandra McGleish)

041 334 5749 (Grant Slessor)

0292 315 820 (Ian McGlerty)

The Housing Department or Environmental Health Department may be able to help you if your house is inside a Housing Action Area (HAA), or is in their future programme of HAAs.

Check with the Environmental Health or Building Control Departments whether they are contemplating serving a closing or demolition order on your property. Your building will have to be in a very poor state of repair before these are likely to be considered.

COSTS	BENEFITS
Actual cost of repair.	The increase in value of your property.
Cost of taking out a loan for legal fees, interest etc.	The effect on the length of life on your property.
Cost of disruption, time off work, temporary housing, redecoration, etc.	The savings from preventing further damage.
Money obtained from grants.	Money obtained from grants.
Any other disadvantages even if you cannot put a cost against them.	Increased amenity and comfort.
	Beneficial effect on insurance premium.
	Greater Safety.

The Value of Repairs

You should expect to pay for repairs to your tenement each year just as you should pay to have a car serviced regularly. If you do not carry out repairs, the value of your property may drop. Building Societies may be willing to lend to potential buyers but they may keep aside a **retention** until repairs are carried out, although this is most commonly done for internal repairs and flat upgrading. Such retentions may stop

Is your building worth repairing?

You may think your property is too old to be worth repairing and that the District Council will be arriving any day to knock your house down. If you think this you will be surprised to hear that in 1989 only 875 houses were demolished in Scotland. If this rate of demolition continues then our houses will have to last an average of 200 years each. People rarely gain financially from

Try to find out whether it is easy to get a Building Society or Local Authority loan in your area. This is an indication of whether your area is going downhill. If this is happening in your area, it is worth forming a Residents Association and asking your local councillor to help you try to reverse the decline. You could think about starting a community repairs scheme or getting the area declared a **Housing Action Area.**

If the District Council does not have any plans to demolish your house then its length of life will depend on your actions and your willingness to keep the property in good shape.

Amenity, Comfort & Safety

It is hard to put a price on the improvements to your living conditions which can arise from some repairs.

Improving the **amenities** in your house can add to its value as well as making it a more pleasant place for you and your family to live in.

Comfort in a house generally refers to warmth. Replacing windows can improve your comfort - but so can just repairing the windows. Sometimes you can put a value on comfort. Supposing that replacing or repairing your windows

means that you use the electric fire less. Suppose you could switch off one bar of the fire for six hours a night for 40 weeks of the year. At approximately 5p per unit of electricity, that would save you £84 per year. This is not much of a saving if you are buying replacement windows at a cost of £1000 or more. However, if you do replace the windows you may find that you are not turning off the fire but you are much warmer.

Safety is something which you must always consider. Under the **Occupiers Liability Act (1960)** you are obliged to keep your property safe for anyone using it, and this does not just apply to people invited in. If a slate falls off your roof and hits someone, you could be sued for negligence. If you have proper building insurance you will probably have some cover against your **occupiers liability.**

The general conditions applying to any insurance usually say that you must take reasonable precautions against loss, damage or accidents. Your cover could be reduced, or totally withdrawn if you do not take such precautions. The cover does not extend to yourself, so, if the slate fell on your head you wouldn't get anything.

In a tenement, you could be held responsible **jointly and severally** with other owners. It may be difficult for you to claim against other owners if you were injured by a fault in the common property - you are as responsible as the other owners when things go wrong. If you can prove that you told other owners of the problem but they refused to do anything, then you may be able to claim against the other owners for your injury.

Repair Priorities

If you have decided to scale down your repair programme, you should follow the basic points set out below.

Its worth remembering though that a good part of the cost of any repair contract is the **prelims** (*preliminaries*)- just getting workmen to the job. You might find some less essential jobs can be done more cheaply if they can be lumped together.

1. Get structural repairs done as quickly as possible. They will be the most expensive repairs but if they are not done, it may not be possible to rectify the fault later.

2. Almost of equal importance to structural repairs are repairs to prevent water coming in. Water can damage all the parts of a building - rotting wood, crumbling stone and mortar and ruining plaster and decorations. If you cannot afford a permanent repair, get a temporary job done and start saving.

3. Some repairs are preventive maintenance - 'a stitch in time saves nine'; these include:

> *external paintwork*
>
> *gutter cleaning*
>
> *mastic around windows*

It's worth getting these jobs done regularly.

4. If its a straight decision betweeen an individual and a common repair, go for the common repair first. The situation may change later and you might not find the rest of the neighbours so keen on repairs another time.

5. Leave the decorative work or other improvements till last. Incomplete or patch repairs can eventually ruin the benefit of internal decorative improvements.

What Adds Value to your Home?

Estate Agents and surveyors all agree that the resale value of a house in sound condition and well maintained is generally better than that of a property with mainly cosmetic improvements. Getting repairs done should therefore be your first priority, especially if all the other owners in the tenement are keen. You may not get a second chance.

The next thing to do is to make a distinction between improvements which will add to the value of your house - in other people's eyes - and improvements which appeal mainly to you as an individual such as installing a cocktail bar or an elaborate stone effect fireplace.

If you are in doubt, you may be able to get some free, but general, advice from a friendly estate agent. You can also pay for specialist advice from a surveyor. This will cost £75 or so but there is no set rate for this kind of advice so shop around.

This is an estimate of what other improvements can add to the value of your house:-

Central Heating: Expect a return of 50-90% of installation cost (gas fired heating does better than other types). Central heating may lead to lower fuel bills as well.

Fitted Kitchens: The more you spend, the less you are likely to get back. Stick to the basics and keep colours neutral.

Double Glazing: May make a house more saleable but you will only recoup 50% of the installation cost.

Fitted Bathroom: Opinions vary on what you will recoup but everyone agrees that it will make a house more desirable.

Insulation: A well insulated home will sell faster and have lower fuel bills.

Decoration: May help to make a house look well maintained and therefor encourage a quicker sale. However more cautious enquirers may think you are trying to cover up faults. The effect on value is likely to be marginal.

Grants

Some grants are still available for repairs and improvements from your local district council. Details of these are given below. The government is planning to change the grant system soon but it looks as though tenement owners will get a better deal than other householders. The new grant rules haven't been decided yet but a lot of the information we give here will still be of help.

Which grant should I apply for?

If you need to install a bath, WC or wash hand basin because the building has never had one or the existing facilities are inadequate, the District Council must give you a **standard amenities grant** (if, of course you meet the rules). Improvement grants may also be available where other work is required, apart from installing standard amenities. If your house is above the **tolerable standard** and has all the standard amenities you may only be able to claim for a **repairs grant**. If you have a lot of work to get done, you may be able to claim the repair grant first and go back later to get the improvement grant as well.

Getting the grant for your house

If you think you may be entitled to a grant, you should first make a quick check with your District Council to see if money is available and how they are using it. We have checked the information in this book but different District Councils interpret the rules in different ways and the rules themselves can change.

When Should You Apply?

You must have grant approval before starting repair work. You should expect to take at least one month to prepare your application. Expect the District Council to take at least one month to process the grant, but ask how long they expect to take. For tenement properties, the District Council may want all the applications to come in at once. The District Council receives its money in April

every year, and the money often runs out at the end of the year. Ideally, you should start asking about grants in November/December, getting approval in April, with work taking place during good summer weather. There is nothing to stop you starting the process at any time of the year however.

There may be a considerable waiting time for grants in some areas. You may need to consider taking temporary measures while waiting for a grant.

Filling in the Forms

Most District Councils ask you to fill in an application form for a grant. You will need to submit **plans** (sometimes several copies) and **quotations** with the application form. Your application won't be accepted unless you have answered all the questions and submitted all the relevant bits of paper. Ask the desk clerk to check that everything is there when you hand in the form. The desk clerk will help you with problem questions but you should have a go at filling in the form yourself first. You should also get written approval for the work from your building society or bank and any tenants living in the flat. You will need to include these letters with your form.

Grant Rates
(as at April 1991)
(Figures from Glasgow District Council, other authorities may have different limits, always check first)

	Limit
1. General improvement grant limit	£12,600

2. Rehabilitation of pre-1914 tenements in housing action areas–

Work carried out by a housing association	£19,700
Otherwork	£17,100

3. Standard Amenities

Fixed Bath orshower	£450
Hot & cold water to bath/shower	£570
Washhand basin	£170
Hot and cold water to wash hand basin	£305
Sink	£450
Hot and cold water supply to sink	£385
Water closet	£680

4. Repairs in association with standard amenities

Per standard amenity when the house has an expected life of ten years or less subject to a maximum total of £1380	£345
When the house has an expected life of more than ten years: (or 50% of approved expenses if that is greater) subject to a maximum total of....	£3450

5. General repairs grant limit	£5,500

6. Repairs to pre-1914 tenements subject to a common repairs scheme	£7,800

7. Means of escape from fire in houses in multiple occupation

Required work	£9,315
Ancillary work	£3,340

You will get grant up to 50% (75% for standard amenities) of the above limits with discretionary 90% grants for hardship and HAAs or Improvement Orders.

Check levels with your local authority.

Grant Rules – The Small Print

There are standard rules laid down by the Government which apply to the various kinds of grants. If you have applied for a grant check all rules relevant to that grant.

KEY

RG   Repair Grant

SA   Standard Amenities Grant

IG   Improvement Grant

DG   Disabled Grant

LPR   Lead Pipe Replacement Grant

HAA   Housing Action Area Grant

ALL   Rules relevant to all grant applications.

When The District Council Must Give Grants

The District Council must give you a grant if they serve an **Improvement Order** (S88, S236 Housing (Scotland) Act 1987).

The District Council are obliged to give grants for properties in **Housing Action Areas,** providing other conditions are met (see text). (HAA)

The Council are obliged to give grants if they serve a **repairs notice**, providing other conditions are met. (RG)

This also applies to houses in multiple occupation, which may also include special notices for the installation of fire escapes.

When You Cannot Get Grants

The District Council cannot approve a grant where the rateable value of the house is too high. The rateable value limit does not apply:

(i) If the house is rented out to someone not in your family (ALL)

(ii) If an improvement order has been served (IG)

(iii) In Housing Action Areas (HAA)

(iv) To disabled improvements and conversions (DG)

(v) To providing standard amenities (SA)

(vi) To some other conversions

(vii) To lead pipe replacement (LPR)

(viii) To business premises getting grant under **Section 108 (S24) Notices**

You cannot get grants if your house was built after 15 June 1964 although in some cases the District Council can change this if they obtain the Secretary of State's permission. The age of the house does not matter if you are applying for a lead pipe replacement grant or for disabled conversions.

You can only get a grant for business premises if you are carrying out common repairs in a tenement under **a Section 108 (S24) Notice.** Check with the council first as they may be prepared to consider your application then place a section notice on the property to allow the grant to be given.

What You Can & Cannot Include In Grants

The cost of VAT and fees can be included in the grant. (ALL)

If you are doing the work yourself, you cannot claim for your own time, only the cost of materials and any labour you employ (provided you employ sub-contractors on a proper contract). (ALL)

District Council grants must be recorded by the Register of Sasines / Land Register. This will be done by the District Council but you (the applicant rather than the owner) will have to pay these charges. (ALL except LPR and commercial owners)

Grant Levels and Payments

You will get a lower rate of improvement grant if your house is above the tolerable standard and has all the standard amenities or you are carrying out conversions. (IG)

If your house has a short life, for example, more than 10 years but less than 30 years, you may still be able to get a full improvement grant. (IG)

There may be a higher rate of improvement grant for pre-1914 tenements in Housing Action Areas. You get a higher rate of grant in Housing Action areas. (ALL)

If you can show hardship, you may get up to 90% grant. (HAA, Improvement grants under Improvement Orders, and houses in multiple occupancy)

You get standard amenities grants regardless of your income. If you can prove hardship in a Housing Action Area, or where an improvement order has been served, you may be able to get a higher level of grant.

Who Can Apply

Any owner can apply for grants. Other people such as tenants can also apply providing they have the owner's consent and the owner agrees to abide by any grant conditions. However this only applies if the owner meets a hardship criteria, thus presently excluding all district council tenants. (ALL)

Timing

You should not start work before you have grant approval (but District Councils may still give you a grant if you have to start in an emergency). (ALL)

The District Council can put a time limit on the grant. (ALL)

A grant should be paid within one month of work being complete, but payment will depend on all the final accounts being completed. (ALL)

After The Grant

For five years after a grant has been paid, the house must continue to be used as a private dwelling house.

If this rule is broken, the District Council can ask you to repay the grant with interest. (ALL)

For five years after the grant has been paid, you must use the house as your main residence and not as a holiday home. You might have to repay the grant with interest if this rule is broken. (ALL)

For five years after the grant has been paid, you must keep the house in a good state of repair. (ALL)

You may sell the house within five years of getting the grant but the new owner will be obliged to meet any conditions laid down. The District Council can increase the eligible expense limit if they have the agreement of the Secretary of State for Scotland. (ALL)

The District Council may allow you to get more grant (up to the limit) if you find that additional work is needed which couldn't be seen at the time of applying for the grant. (ALL)

In Housing Action Areas, a standard is specified.

You may be able to get a grant for work above this standard. The Council can give you part payments of improvement grant to allow you to carry out common repairs at a different time from improvement. (HAA)

New Grant Rules

The government have been talking for some time about changing the rules governing grants. If the changes are introduced, they should be well publicised before they come into force. The main changes which have been proposed are:-

You will only get a grant if your household income is below a certain level. The rateable value of your house will no longer matter.

There will only be one mandatory (compulsory) grant available. This will cover the work required if your house falls below a tolerable standard. There will be discretionary grants available if your house needs work above this standard.

There will be special rules governing tenements. If some owners qualify for grants, others may get assistance too.

The discretionary grant may have to be repaid if the house is sold within a certain time.

In Housing Action Areas it will be possible to phase improvement and repair work rather than to do all the work at once.

While these changes may make some sections of this chapter outdated, the advice given in this chapter will still be useful.

Types of Grants Available

Local District Councils can give grants to repair and improve properties. If you live in a **listed** or **historic building** you may be able to get extra grants (see your local Planning Department about this). The grants available at the time of writing are:-

The Standard Amenities Grant

This grant is to help people installing bathrooms, kitchen sink, hot and cold water, WCs, wash hand basins etc., for the first time. It is not a big grant but it is mandatory. This means that if you meet the rules, the District Council must give you the grant - regardless of whether it has money available.

If you get a standard amenities grant you may be able to get grant for up to £2500 of repair works.

If you apply for a standard amenities grant, the District Council can insist that you install **all** the amenities if the house has a life of more than ten years.

The Improvement Grant

This grant is for major improvement works. It will include such items as upgrading the electrics, DPC installation, kitchen improvements, major internal alterations. It includes the items covered under standard amenities but is more wide ranging. This grant is discretionary - that is, the Council does not have to

give it to you if it does not have the money, although there are limited exceptions to this. If your house lies within a **Housing Action Area** then, depending on your circumstances, you will be entitled to a higher level of grant. You can also use these grants for converting large properties into smaller ones (see also Disabled Conversions and Improvements).

If you get an improvement grant, the Council can demand that you carry out all the work necessary to bring the house up to the following standard, even if this puts the cost of work over the eligible expense limit:-

The house must be in a good state of repair and free from damp.

The house must have all the standard amenities for exclusive use of the occupants (not shared). You should also have a good hot and cold water supply and drainage system.

The house must have satisfactory ventilation, lighting, heating and food preparation facilities.

The house must have a future life of 30 years.

You can only get an improvement grant **once** unless you got the grant more than ten years ago or you didn't get the full improvement grant first time.

You may be allowed to spend half the improvement grant on repairs. If you are

Approaching the Council for a Grant

Important

Get a note of the name of the person you are talking to.

Take a note of the meeting, write all the answers down at the time or soon after. If appropriate, write back and confirm the points agreed.

Checklist of Questions to Ask

1. Is there money available for discretionary grants this year? *(The District Council year runs from April 1 to March 31).* If there is not enough money for this year, can you be placed on the waiting list?

2. Does the District Council restrict grants to certain groups? *For example, to houses in Housing Action Areas; houses below a certain rateable value or age; or applicants whose income falls below a certain level.*

3. Will you need to bring your property up to a certain standard in order to get a grant?

4. How long does the grant procedure take?

5. Will the District Council make you take the lowest estimate?

6. Will the District Council give you a grant on any additional works which crop up during the contract?

7. Do you need to use an architect or surveyor for grant work?

8. Can the District Council recommend architects or contractors?

9. Will the District Council inspect your house and give you the results of their survey?

10. Will the District Council inspect the work at the end?

11. Will the District Council issue a compulsory repairs notice if you can't get owners to agree?

12. Does the District Council operate **agency agreements?**

13. Will the District Council make grants available to shops or other businesses in your block?

14. Does the District Council give loans for repairs? Other neighbours may need this information even if you don't.

15. If you need rehousing during the work, will the District Council give you a decant house or give you financial help for moving?

16. Are there any other rules you should know about?

carrying out improvement work but need to spend more than half the cost on repairs, the District Council may give you some of the money as a repairs grant providing the two grants together don't exceed the eligible expense limit for improvement grants.

The Repairs Grant

This grant is discretionary, unless the works are part of a **Section 108 repair notice.** This grant also covers replacement of lead pipes to your drinking tap.

The District Council will usually give repair grants for work to the common fabric of tenements, the roof, gutters, downpipes, stonework, rotwork and some structural work. If the work is of a serious structural nature then they may offer improvement grants (ask your District Council first).

The District Council does not have to take into account your earnings when giving a grant for work which is of a major structural nature. If repairs are less serious the District Council has to be sure it would not cause you hardship if you paid the full cost of the repairs.

You may get a repairs grant even if you have previously had an improvement grant, but it is up to the District Council to decide.

Commercial owners will also be able to obtain a repair grant if a section 108 notice is placed on the property.

Lead Pipe Replacement Grant

If you are seeking a lead pipe replacement grant you may have to get tests done on the water to show that the lead content is above the EEC recommended level of 0.1mg/litre of water. However this is usually not required.

The grant allows you to include work which involves: the by-passing, re-lining or replacement of lead lined tanks, the replacement or by-passing of lead piping within the house or its curtilage.

Strictly speaking, you can only get grant for pipes leading to your drinking tap. As

people sometimes fill their kettle and pans from the hot water tap, some Councils may allow the grant to cover all lead pipe replacement.

You won't get a lead pipe replacement grant if the house does not provide satisfactory housing accommodation in the opinion of the District Council.

Disabled Conversions and Adaptions

You can get an improvement grant for making a house suitable for the use of a disabled person. The grant will allow you to put in a second toilet or bathroom if the disabled person cannot get to the existing WC. It should also cover major adaptations such as hoists. Some other

rules are ignored as well. The grant does not cover minor alterations such as hand rails, or supplying equipment. For these you will need to see your Social Work Department.

Environmental Grants

Some councils give grants for environmental improvements including work to backcourts, stone cleaning, replacing binstores, fencing, landscaping and outdoor drying facilities. You are not normally asked to make a contribution to this work providing you keep to a cost limit. These are often the first grants to be cut back on when money is short but, if you live in an Housing Action Area you could make a good case for getting this grant for the whole street.

Insulation Grants

You can get help with draught-proofing and loft insulation through the **Home Energy Efficiency Scheme.** To qualify you have to be in receipt of benefit such as Income Support, Housing Benefit, Family Credit or Community Charge Relief. To apply you will need to contact your local Network Installer (address below or from your local District Council). The 'Network Installer' will check your benefits and then give you a quotation for the work. You can get work done up to £121 for draught-proofing (you pay the first £7.50); up to £188 for loft insulation (you pay the first £10.70). You can do the work yourself but the grant levels are then lower. You can also get the work done through an accredited contractor (check with Energy Action Scotland).

Apply to the E.A.G.A. office in Newcastle:

Energy Action Grant Agency
Bank Chambers
9-17 Collingwood Street
NEWCASTLE UPON TYNE
NE99 1NG
Tel: 091 2301830

Energy Action Scotland
21 West Nile Street
GLASGOW
G1 2PJ
Tel: 041 226 3064

Other Sources of Finance

The current lack of finance for grants in some areas is making it important for property owners to consider other ways of raising finance. This chapter looks at other ways of raising money to pay for repairs: loans, Department of Social Security grants.

If you have to raise money for repairs by saving, remember that all the time you are saving, your repair will be getting worse and more expensive. You may never save enough in time to correct the repair.

Help from Social Security

If your income is low, you may be able to supplement it with state benefits. If you need help in claiming, many local groups exist to give advice to people in your position. Look for claimants unions, community advice and information centres, and Citizens Advice Bureaux. Many social work offices also have Welfare Rights Officers. You will probably need to make an appointment but you should get thorough advice and useful help in complicated cases. You can also make a free call, anonymously, to the Department of Social Security at 0800 666 555.

Income Support and Housing Costs

The **Income Support Scheme** replaced **Supplementary Benefit** from April 1988. You are entitled to claim Income Support if you are not working or if you receive other state benefits. Under the Income Support Scheme, you will get a fixed amount of money per week, depending on the number of people in your family, their age and disabilities. You are expected to pay most of your housing costs from this weekly sum. This includes the first 20% of your poll tax, all water rates, insurance and the cost of minor repairs. If you receive unemployment or invalidity benefit, the amount you pay in mortgage interest may make you eligible for Income Support even if at first sight you don't appear to be eligible.

If you are receiving Income Support, you can claim help for carrying out repairs from the **Social Fund**. You are not expected to be able to budget for high cost repairs when you are on

Income Support but it will be up to the Department of Social Security to decide what exactly is a costly repair. You will be expected to try and get financial help from elsewhere, *for example, seeking a loan or grant from a charity, or from friends and relatives,* before you apply to the Social Fund. If you do succeed in getting money from the Social Fund, it will be in the form of a loan and you will be expected to make weekly repayments on the loan. This could be as high as £10 per week.

Income Support and Mortgage Interest

If you have a **capital repayment mortgage**, each payment you make to the building society will be partly an interest payment on the loan and partly a capital repayment of the loan. If you receive **Income Support** you will be able to get help with paying the interest part of the loan.

If you have an **endowment mortgage**, the Department of Social Security will calculate what your interest charges would be if you had a capital repayment mortgage and give you that amount of money. You will not get help to pay the capital or the insurance parts of your mortgage. The general rule is that you will receive half of your interest charges for the first sixteen weeks you are on Income Support and the full amount after that. There are exceptions to the sixteen week rule. These are:

You or your partner are over 60.

You are claiming Income Support after a break of less than eight weeks.

You have recently separated from your partner who was claiming Income Support.

In these cases, you may get all of your mortgage interest paid immediately or in less than 16 weeks.

The Department of Social Security calculates how much Income Support you are entitled to by comparing what they think you need to live on with your income and paying you the difference between these two figures. It is possible that when you first claim income support you will not receive any payments because only 50% of your mortgage interest costs are involved in

the Department of Social Security's calculations.

If you do not get Income Support initially, you should claim again for Income Support, sixteen to twenty weeks after your first claim. If in doubt, claim earlier. In this case, you will get all of your interest charges paid immediately your claim for Income Support is accepted.

You won't get help with interest payments on any **arrears** which build up except where these are built up during the time you only got half of your interest charges paid.

You may also claim for Income Support to pay the **interest on a loan** taken out to pay for repairs. Beware, however, if you are receiving any transitional protection. When you claim for help with interest payments for a new loan, you will be treated as making a new claim and so lose transitional protection. If this is more than the weekly interest payments on the loan, you could lose out. Take welfare rights advice.

Loans and Department of Social Security

If you cannot get help from the Department of Social Security Social Fund, you may need to get another loan. Explain, when asking for a loan, that you will be able to get the interest charges only paid by Department of Social Security. Include, in the loan you ask for, the cost of legal fees in arranging the loan as these will not be paid by Department of Social Security. If you have savings of more than £300, the excess savings will be deducted from the loan when the Department of Social Security calculate interest.

You can claim interest payments on taking out loans for the following:

Major structural repairs.

Repainting, rewiring etc.

Installing standard amenities.

Damp proofing and extra ventilation or lighting measures.

Electrical work.

Drainage works.

Provision of heating.

Insulation.

Other 'reasonable' improvements (these can be wide ranging for elderly and disabled claimants).

Again, get Department of Social Security agreement in writing before committing yourself to the loan. The rules for Department of Social Security help for mortgages apply.

What To Do If You Have A Mortgage and Lose Your Job

Apart from signing on and making your claim for Income Support you should immediately **tell your Building Society** (and insurance company if you have an endowment mortgage) so that they can make changes in your loan. Most Building Societies and other lenders will extend the length of your loan and put you on to an interest only payment basis.

Exactly what the Building Society will do depends on how much you owe and the value of your house. If you have only recently taken out a mortgage and it was for 95% or 100% of the house cost, the Building Society may suggest that you sell your house, perhaps allowing you to buy something cheaper. If you have had your mortgage for some time, have made plenty of payments and the value of your house has increased, then the building society is generally able to be much more flexible and extend your loan for a longer period.

If you have an endowment mortgage and can't manage the insurance payments, the insurance policy will lapse. If you have only recently taken out the policy, you may forfeit all your payments. Otherwise, your policy may have a **surrender value** which may be paid to the building society, reducing the amount of your outstanding loan. When you start working again you'll need to take out a new insurance policy.

If none of these arrangements work out for you then, as a last ditch effort, you could consider taking in a lodger. Generally speaking, the income you get from letting out part of your house will not count against Income Support if you use the money to pay off your mortgage. The rules can be complicated however, so check with an adviser before you go ahead.

Where To Go For Loans And Mortgages

Building Societies
Building Societies are the largest lenders of money for house purchase, home improvement and repairs. You will usually only get a loan from a building society if they can register the first charge on a property, that is, there is no other mortgage outstanding on a property. Building Society mortgages are among the cheapest (in terms of interest to be paid) and they are normally given over a longer term so that your monthly repayments are lower. Different Societies have different policies and slightly different interest rates. It's worth shopping around.

Banks
Banks are a good source of loans and mortgages. Some banks offer special home improvement loans. Check the rates carefully - extending your existing mortgage may be cheaper.

Insurance Companies
Some insurance companies are prepared to give second mortgages but remember that these are more expensive than first mortgages. You might also need to take out more life insurance cover to get a loan from an insurance company.

Local Authorities
Your District Council can give loans for repairs or improvements and they may be more flexible than a building society or bank. If the District Council is forcing you to carry out repairs or improvements, such as by serving a **compulsory repairs notice** or declaring a **Housing Action Area**, then the District Council must give you a loan, providing you can repay the loan and providing the value of the house is greater than the loan. The District Council can give loans without having first charge of the property and if they don't have enough money themselves they can act as a guarantor for a building society loan.

Finance Companies
Finance company loans can be very expensive. They should be considered a last resort. They may charge a very high rate of interest but try to disguise the actual amount from you. If you don't pay tax, you won't get any help from the government for this kind of loan. They

will probably only give you a loan over a short period so you will also have to repay a lot of capital each month. Some finance companies charge a fee for arranging loans (which could be one tenth of your loan) which will be added to your loan. This fee is sometimes called a **guarantee** or **indemnity premium.**

Sometimes, a clause is also written into the loan agreement which makes you pay even more money if you want to repay the loan early. You must read the small print, ask what the **Annual Percentage Rate (APR)** is, check on fees and ask what your final monthly repayment is before you sign anything. Work out what your monthly payments will add up to paid over the years of the loan. Find out what will happen if you miss repayments.

If you find yourself in difficulties, you may get protection under the Consumer Credit Act 1974. This Act is put into action by your Regional Council's Consumer Standards Office. If you have problems see them first. If you cannot get any resolution to the problem through the Consumer Standards Office then you will need to consult a solicitor. Your solicitor may be able to get the court to change the terms of the loan.

Credit Unions
If you are a member of a credit union, you may be able to get a loan from them. To set up a credit union, or find if there is one in your area contact one of the following:

National Federation of Savings and Co-operative Credit Unions

1st Floor, Jacobs Well,

BRADFORD BD1 5KW

TEL: 0274 75350

Association of British Credit Unions Ltd.

Unit 307, Westminster Business Square,

339 Kennington Lane

LONDON SE11 5QY

TEL: 071 502 2626

Strathclyde Credit Union Development Agency

95 Morrison Street

GLASGOW G5

TEL: 041 429 8602

Arranging a New Loan

To get a loan, you will need to show:-

That you can afford the repayments.
You will need to check the interest rates
charged by different lenders. Remember
to take into account tax relief - if you
don't pay tax, ask about **MIRAS**
schemes or **option mortgages.** If you are
on Income Support you should ask about
getting the interest paid by Department
of Social Security. If you are not
working, the interest payments on the
loan may push up your needs so that you
are entitled to income support.

**That the value of your house after
repairs or improvements is greater
than the loan you need.** You may need
to pay a building society to value your
house. Don't forget that you may need to
pay legal fees for arranging the loan. Try
to get these added to the cost of the loan.

You will also need to show your lender:

Repairs estimates.

*A note of the grant you will get from
the Council.*

Details of your income.

If you are not working you will need to
show that the Department of Social
Security will pay your interest. It may
help to make a note of your outgoings
and other commitments, in order to show
that you do have money left over for a
monthly repayment.
If you are having trouble finding a
reputable lender ask your local District
Council for help, their **Home Loans**
section may be able to suggest some
sources. The **Social Work Department**
may be able to offer emergency loans.
Sometimes a charity will give help. You
should ask to see The Directory of Grant
Making Trusts in your local library. You
may get help because of your age or
disability, your occupation, where you
live or were born, because you have
children or because you served in the
Forces.

Extending Your Mortgage

Extending your mortgage is one of the
better ways of paying for repairs. Before
extending your mortgage, you will need
to consider:-

*How you are going to meet the extra
payments. If you cannot afford to pay
more, you may be able to get your
lender to extend the term of the loan
so that you pay over a longer period.*

*Whether the value of your house
(after the work has been done) is
greater than your extended
mortgage.*

You will certainly need to get your house
valued and you will need to pay for a
building society surveyor. The surveyor
will want to know what repairs or
improvements you are planning.

If you are having difficulty extending
your mortgage:-

*Check whether you can get other
help - see Department of Social
Security.*

*See your District Council's Home
Loans Section - they may be able to
help by referring you to other
building societies or giving you a
loan themselves.*

*Consider transferring your entire
loan to another building society that
will give an additional top up. Ask
your neighbours' building societies
as they will already have an interest
in your tenement and your area.*

Don't forget to ask your lender to include
the cost of legal fees in your loan.

Further reading:

*Rights Guide for Home Owners. London:
Child Poverty Action Group.* This is
usually available at your local library and
Citizens Advice Bureau. It has a lot of
useful advice on how to negotiate with
Building Societies and other lenders.

Loans for the Elderly

There are two schemes especially
designed for old people on low incomes.

The Maturity Loan

A maturity loan is one where the
borrower repays only the interest on the
loan. When the borrower sells or dies,
the capital of the loan is repaid from the
sale price of the house. Local Authorities
and some Building Societies offer these
loans. The elderly person may get the
interest paid if they are entitled to extra
help from the Department of Social
Security. The department will, however,
treat them as making a new claim and
they might then lose any transitional
protection they have. Get welfare rights
advice on this before making your claim.
In exceptional circumstances you may
get **Maturity Loans** with **rolled up**
interest. This means that the interest is
added to the capital and repaid with the
capital eventually.

Home Income Plans

These are offered by a few insurance
companies and financial organisations.
With these schemes, you take out a fixed
interest mortgage on your house (the
mortgage can be anything up to
£30,000). You then buy an annuity with
the mortgage money. An annuity gives
you a regular income for as long as you
(and your spouse) live. Your relatives
will probably have to sell your house
when you die to repay the mortgage.

I SEE JIM GOT HIS MORTGAGE EXTENDED!

The interest payments for the mortgage are deducted before you are paid your annuity.

Two publications from Age Concern called *Using your Home as Capital and Raising an Income from your Home,* give more details of some of these schemes. The value of these schemes depends on your circumstances. To be eligible you will need to be at least 65, (the combined ages of a couple should be 130). Your property must be owned outright. As these schemes affect your entitlement to benefit, and as they are fairly complicated, it is essential that you see a financial advisor before embarking on this kind of scheme.

Old people often fear 'going into debt'. They may have additional fears with both these loan schemes because they feel they will be leaving a debt to their relatives. There are very few of us who would wish to see our elderly parents living in misery just so they can pass on their home to their children clear of debt. With a **maturity loan**, there is a chance that the house will increase in value more than the debt and sons and daughters may get a better asset than if the house had been left unrepaired.

Further Reading:

Home Income Schemes
Which, January 1992.

C. Hinton (1991) Using Your Home as Capital. Age Concern: London. Updated edition price £3.50.

Raising Income From Your Home. Age Concern: 54a Fountainbridge. Edinburgh EH3 9PT. Free leaflet.

For advice, information and counselling phone: Age Concern information line: 031 228 5467.

36

Mortgage Arrears

If you have fallen into arrears with your mortgage because of unemployment, sickness, etc., you should go and see your lender immediately. Don't ignore the problem because it will only get worse. If you get proper help in reasonable time, you can manage debts. If there is a charge over your property, your lender can make you sell the house but they will need to get a Court Order first. This order takes some time to arrange. Your lender will write to you (probably twice) asking what action you intend to take to reduce the debt. Your lender will then see solicitors who may also write to you before issuing a summons. The summons will tell you when the Court hearing will be. If proceedings get as far as court, one of three things could happen:-

The court will dismiss the action if you have now managed to pay the debt and appear likely to continue paying.

The court can issue a suspended possession order giving you a further chance to pay off the debt.

The court will agree to the possession order.

Even after the possession order has been granted, the lender will still need to get an Eviction Order.

During this fairly lengthy procedure, you should do everything you can to retrieve the situation. Try any or all of the following:-

Pay whatever you can every month. Do not let the debt get bigger than is absolutely necessary.

Check ways of increasing your income, check you are getting enough benefits or consider taking in a lodger.

Write to your lender explaining your situation and what you are doing to get out of debt. You may be able to re-negotiate your loan so that it is spread over a longer period.

Ask if your mortgage can be changed so that you pay interest only.

If you are not working, check the section on Income Support. You may be able to make a back dated claim.

See the local Social Work Department if you have children under 16.

Get advice from local advice centres, Citizen Advice Bureaux, etc.

Taking Legal Action

Taking legal action is, literally, the last thing you should do, unless someone is taking legal action against you. Sometimes the threat of legal action is enough for people to focus their minds on a solution to the problem. The following chapter sets down some of the actions you will need to take.

Before you See a Lawyer

This book sets down several courses of action which may resolve the dispute. If you have tried everything then check the following:

Talk your problem over with someone a bit more knowledgeable - a **local advice centre** would be a good place to start. Your adviser could tell you whether or not it's likely to be worth going ahead. You may be legally right but it might cost a lot to prove it. The advice centre could also put you in touch with a lawyer who specialises in your kind of problem. The advice centre may also be able to tell you if you are entitled to **legal aid** or advice and assistance.

A conciliation or arbitration scheme may be in operation. Use it if you have problems with someone who is a member of a trade or professional association.

Read about your problem and jot down notes about your case. If you need to go and see someone, you will get a lot more out of it if you know what they are talking about and if you can give them all the relevant information quickly and clearly.

Check other Sources of Free Help:

Your local Consumer Advice Centre or Trading Department might take up your case - a visit from one of their officers might sort out an offending tradesman immediately and at no cost to you.

Ask at a Housing Aid Centre or Rent Tribunal. Again, they could put pressure on the offending party.

Ask your employer for help - there may be a solicitor working for your firm who could give you some free advice.

Conciliation and Arbitration Procedure

Try to find out if the person or firm is a member of a trade professional association. There may be initials after the person's name, a logo on the letterhead or bill or a certificate displayed in the shop or office.

Ask that Association for details of any scheme that they operate.

Start off with the conciliation procedure which is designed to bring two disputing parties together to get some agreement.

Ask for details of the arbitration procedure. It might be cheaper than taking legal action.

Check the following:

Is the scheme free or is there a cost?

How long will it take?

Is there a cash limit?

How much time and trouble will be involved?

Will you need to travel to a hearing?

Will you forfeit any other legal rights by taking this action?

Visit a Solicitor

First find your solicitor. You may go by personal recommendation or you may wish to check the *Directory of General Services provided by Solicitors or the Solicitors Referral Lists* to find a specialist. These publications can be seen at Libraries, Citizens Advice Bureaux, Sheriff Courts, etc. These lists will also say whether the Solicitor will offer a fixed fee interview or legal aid. The **Scottish Legal Action Group** (or the Legal Services Agency in Glasgow) may help find a solicitor for groups who are taking up a campaign.

The Fixed Fee Interview

Many solicitors offer this service - you will probably get a half hour of advice for a modest fee (ask before you begin the interview!). This is an initial interview to help identify a person's problem and what can be done about it. At this stage you should also ask the solicitor for an estimate of what legal action will cost and whether you are eligible for **Legal Aid**.

Action in the Sheriff Court

If you are involved in legal action, you will probably find yourself in the Sheriff Court. The Court of Session is used for divorce cases, appeals and more expensive or complicated forms of legal action.

The Sheriff Court will deal with the following kinds of legal action:

Small Claims Procedure

The scheme is designed to deal with claims under £750. Both parties will be asked to come and tell their story to the Court. You will not need legal representation and if you lose you will only have to pay the other side's travel and witness costs on top of the amount of the claim (get details from the Sheriff Clerk's Office or the CAB).

The Summary Cause Procedure

This is used for debts or damages between £750 and £1500, claims in connection with defective goods and actions over the tenancy of a property. This is the only kind of action which is designed to be used without legal assistance. If someone takes action against you, it is most likely to be a Summary Cause action.

The Summary Petition

This is a kind of action designed to operate fairly quickly. You would use it to stop something happening which you didn't like *(for example, appealing against a Section 108 Notice see page 38).*

The Ordinary Action

You would use this if you wanted a judgement made about something which is unclear *(for example, an argument over a common repair)* or if you wanted to make a claim over £1500.

How Legal Expenses are Calculated

Court work undertaken by solicitors (but not advocates) is charged for at rates specified by the court. There will not be much difference between solicitors but it will be difficult to get an estimate of likely costs simply because it is so hard to predict how long the court case will last. Ask your solicitor anyway.

Non court work is charged for on the basis of expenses plus fees plus VAT. The fee rate will differ from solicitor to solicitor but will take into account:-

how much time it takes.

the amount involved.

the importance of the case to you.

the degree of skill involved.

the complexity or novelty of the case (this means it can pay to go to a specialist).

the number of documents involved.

where and how the work is done.

The **Law Society** publishes a guide to the maximum amount solicitors should charge for certain types of work but these scales can be reduced or increased. Time is often charged for at about £60-£100 per hour depending on the seniority of the solicitor involved.

Legal Expenses Insurance

It is now possible to get insurance against having to take legal action. It costs about £60-£120 per annum but does not cover house purchase or divorce actions. You will only be covered for actions which could only reasonably have been discovered during the period you had insurance.

Taking Legal Action Where More than One Person is Involved

If all or most of the owners of a tenement are involved, *for example by someone refusing to pay for a common repair or a tradesman not doing work properly,* then try to get someone who is entitled to **legal aid** or who has legal expenses insurance to take action first. They may either take action for the whole sum involved *(this is appropriate if you are 'wholly and severally' responsible for payment of the bill as co-owners)* or they can take action on their own behalf only. If they win, the debtor or tradesman may give in and sort things out with everyone

in the tenement and it will be much cheaper to launch other actions on the back of the first.

If none of your neighbours wants to take legal action, it could just be that they don't have your confidence to try something new. It will be helpful if you all act together, but don't let lack of support stop you if you are convinced you are right.

Someone Takes Legal Action Against You

Don't bury your head in the sand. If you do nothing you will certainly lose the case and be forced to pay the other side's legal and court expenses on top of the bill. You must act immediately. You could ask for help at a local advice centre or Citizens Advice Bureau. You could also see a solicitor on a fixed fee interview. Don't forget, if you win the case, you will be awarded your expenses (look at the sections on Summary Cause and Appealing against a Notice).

Further Reading

You and Your Rights in Scotland. Readers Digest (1984).

Where to go for Legal Advice. Which. April 1991. Consumers Association.

The Legal System of Scotland, HMSO (1975).

Legal Expenses Insurance, Which. April 1991.

Various leaflets on specific topics are obtainable from

Sheriff Courts or the Citizens Advice Bureau.

Negligence

If you do not have a contract, you may be able to sue for negligence. To prove negligence, you must prove:-

the existence of a duty of care.

a breach of that duty of care.

that you have suffered loss or damage.

The **duty of care** depends on who you are talking about. In the case of a professional, such as an architect, the duty of care is defined by the standards of the profession and not what you would expect from the ordinary man in the street. A professional person is an expert by training and experience and must be as careful and skillful as the average member of their profession.

If you are looking for damages, you must show that the negligent person could reasonably have foreseen that the damage would occur. *For instance, the builder has stripped your roof and while he is not working on it over the weekend, has covered the roof with tarpaulins. Over the weekend, gale force 9 windows blow up and blow the tarpaulin off the roof. The builder has not been negligent as he took precautions against foreseeable rain but could not foresee the gale force winds.*

Architects and Negligence

An architect is expected to have a working knowledge of the law as it affects the profession. For instance, you could expect the architect to tell you that you need Planning Permission or a Building Warrant or anything else which is reasonably forseeable that you could not be expected to know about. Failure to inform you under such circumstances would probably be negligent.

Appealing Against a Notice

Many **notices** served by District Councils say that you may appeal to the **Sheriff Court.** You will need to use the **Summary Application Procedure.** It costs £18 to lodge the **Initial Writ** at a Sheriff Court. You may also need to pay legal expenses and other fees in addition to this. You should first see the council and discuss the serving of the notice with them. You could get independent advice from a building surveyor, architect, solicitor, or advice

agency, whichever seems to be appropriate.

If you decide to go ahead with the appeal, you must get your Initial Writ lodged at the Sheriff Court within the time limit set out in the notice. The Sheriff Court will issue a warrant which you will then serve on the council. You will only need to set out the bare bones of your argument - you will be able to give more details at the court hearing which will be in some 3-4 weeks' time.

If another professional is employed by you or your architect you may have trouble deciding who is negligent. *For instance, the architect could be expected to rely on a consultant structural engineer who has been employed and the engineer could be held negligent if, say, the structure was unsound.* In this kind of case you can expect guidance from the architect.

A much more difficult problem to solve is one where there may be negligence over **inspection** and a **clerk of works** has been employed. It is accepted that an architect will visit the site about once a week and the architect is expected "to inspect generally the quality and progress of the works". The architect does not and cannot check for conformity of everything they see. If you want higher levels of inspection, this can be negotiated with the architect. Always remember that under the contract for carrying out the works, responsibility for supervision and maintenance of quality is the contractor's alone.

The clerk of works is expected to check the workmanship and details of the work at frequent intervals, usually daily. If the architect employs the clerk of works, then the architect can be held responsible for him. If you employ the clerk of works then it is the clerk of works who may be sued for negligence, rather than the architect *(but you might end up suing both the architect and the clerk of works and letting the court decide who exactly was responsible).*

You could expect the architect to point out obvious shortcomings in the work of the other professionals employed. If the architect agrees to do the work that could be done by another professional, you should still expect him to do a competent job.

On the subject of costs, you can expect your architect to advise you whether the amount you have allowed for a job is sufficient to do all the work you intend.

There is a court case dating back to 1876 where the lowest tender received on a contract was 50% more than the client had told the architect he was prepared to spend. The court said that the client was not obliged to pay the architect's fees as the architect's conduct had suggested that it was worth going ahead with the job.

When it comes to issuing **certificates** during a contract it is much more difficult to say that an architect has been negligent because the role played is one of an impartial judge between the client and the employer.

Problems With Architects?

The RIAS offer several procedures which can be taken up where you have a disagreement with your architect.

The Second Opinion

If you have serious doubts over a professional judgement or advice given by your architect, you may consider obtaining a second opinion. Normally that opinion would have to be paid for by the client unless the architect agrees otherwise. The RIAS can advise on such an appointment.

Conciliation procedure

This is aimed at bringing the two parties together to resolve a situation where no money is to change hands. In general, complaints of failure by the architect which the RIAS are able to undertake would cover the following subjects:

1. Disgraceful Conduct

2. Conflict of interest

3. Failure to have Client's approval

4. Failure to adhere to budget (with reference to Quantity Surveyor's advice).

5. Failure to observe the procedures laid down in Part III of the Architect's Appointment, or in the Conditions of the Standard Form of Agreement for the Appointment of an Architect (SFA 92).

The RIAS will send you a form to fill in which asks for details of the job and your problem. This form is then sent to the architect who is asked to give his or her side of the problem. All the papers are then put before an investigation committee who make a judgement on the problem. If there has been disgraceful conduct, the sanction of suspension or expulsion from the RIAS may be used. You cannot use this procedure if you are looking for damages.

The RIAS is currently considering what help it can give informally as a conciliation service and hopes in future to be able to nominate mediators to assist parties in dispute towards an amiable settlement.

Arbitration

This is more suitable where you want to get some financial redress. Under the standard form of appointment, disputes that cannot be settled by a second opinion or conciliation must, unless the parties agree otherwise, go to arbitration. Both you and the architect must agree to the choice of arbiter. The RIAS can nominate architect arbiters but if you can't agree you can ask the Dean of the Faculty of Advocates to appoint someone.

Arbitration can be undertaken simply on 'documents only' - or with a hearing, with or without legal agents - so it can be adjusted to suit your case. You may wish to bring an expert witness to act on your behalf, again the RIAS can nominate suitable senior architects to act in this capacity. (Indeed you may decide not to proceed with arbitration after you talked to the expert witness). An expert witness will often have a lot more authority in saying that something has gone wrong (which your architect may deny or at least try and play down). Don't forget that the expert witness will be able to claim fees - ask what they will be before going ahead. Your arbiter will probably be an architect who you will expect to make a quick, sensible and business-like decision. The arbiter may appoint legal assistance where this becomes desirable. You must make sure the arbiter is clear about what their job is: the matters you want considered, whether you want damages etc. The arbiter will not go outside these guidelines. You can expect the arbiter not to see the architect or to visit the site without your being there (and vice versa). You will probably need to agree that the arbiter's decision will be final and legally binding.

If you go to Court you will probably be considering a case of Negligence or Breach of Contract.

If your architect is not in the RIAS or RIBA you will have to go direct to the Architects Registration Council UK. They have their own discipline panel who have the powers to remove architects from the register.

Breach of Contract

If you have a contract with someone and they do not keep to that contract you may be able to sue for breach of contract, although you will need to show that you have suffered financial loss. Often the contract is very complicated and may refer to other documents such as Conditions of Engagement issued by a professional body. You should first of all take up your complaint with the other party. If you get no satisfaction from them, you could contact the appropriate professional body and ask for their help. They won't make a judgement for either side but may be able to advise you on what is generally considered reasonable. If you still want to take legal action, you should consider seeing a solicitor.

The Council and Negligence

A council can be sued for the actions of an employee. You should be able to rely on the council to do a job properly. Do not forget that you must show loss or damages resulting from their actions.

The actions of a council officer may be the basis for an action of negligence.

For example, in the English case of **Ditton v Bognor Regis Urban District Council,** *a house was built on a disused rubbish tip. The house started to subside because of inadequate foundations. There was a council by-law that said houses had to have adequate foundations and that the council building inspector should see that they complied with the by-law. Mrs Ditton, who was the second owner of the house, took the council to court because the inspector had negligently approved the foundations.*

The point to note here is that the second owner sued the council. The second owner had suffered the loss and the council could easily foresee that the house would have subsequent owners.

Taking Action Against the Council

Apart from negligence, you can also sue a council for not abiding by its statutory duty (for example, the council has a statutory duty to enforce the Planning Acts). You can also argue that a council is acting beyond its power - this is known as **ultra vires** (for example, the council closes off a street without going through the necessary consultation procedures).

You may also be able to claim maladministration, and thus seek the local Ombudsman's assistance (see p 52).

Further Reading

Professional Negligence. Underwood & Holt, Formal Publishing (1981). Hudsons Building Contracts.

40

A Tenement Alphabet

Absentee Owners

Tracing Absentee Owners
If the flat is let, the tenants should have the owner's name and address in their rent book. If they haven't, check the **Property Register** held by your Regional Council. This replaces the Valuation Roll since the introduction of the community charge. The property register will be kept by the **Regional Assessor.** You may get the landlord's agent's name and address only. The agent should tell you who the landlord is and pass on a letter to the landlord. If you don't trust the agent, or can't get any further information (perhaps the house is empty), you could write to the *Registers of Scotland, Meadowbank House, London Road, Edinburgh.* Give details of the flat position (try and give a compass reference as well), when it was last sold and the previous owner's name (if you know it).

The Registrar will then write back telling you what the search fee and photocopying costs are and when you send the search fee, you will be sent the appropriate details. The current search fee is £6.81 and photocopies are charged at 35p. per page.

For example, a group of owners involved in a tenement were trying to trace the owner of an empty flat in their close.

They traced the owner through the Registrar - he turned out to be a Chinaman who said he had used the flat as a bet in a Mah Jong game and lost. However, no conveyance had taken place and so he was still legally the owner.

Agency Agreements
Some District Councils will offer owners undertaking building work an **agency agreement.** The District Council is allowed to do this under the **Housing (Scotland) Act 1987.** The District Council becomes the owners' agent, hiring architects or surveyors, signing the contract and paying the builder. The District sends the final bill to the owners, having deducted the amount of grant the District Council would be due to pay the owner. The

main advantages of this type of agreement are:-

> *Owners don't have the responsibility of taking on a contractor, architects etc.*

> *Under a normal contract, the builder would be paid regularly and the owner would only get grant some weeks after the job had finished. The owner might need to take out a bridging loan to cover the payments. Under an agency agreement, the District Council covers all the payments.*

The main disadvantages are:-

> *The District Council may add on a charge to cover their administration and interest costs.*

> *The District Council can tell the contractor to do extra work which the owners don't want and the owners will still have to pay for it.*

If the District Council offers an agency agreement you will be asked to sign a legal document. Read it carefully so you know exactly where you stand. Some District Councils such as Edinburgh and Glasgow do not like using agency agreements. In Edinburgh's case, this is a decision that was made following a case where a builder had gone bankrupt and residents had held the District Council responsible (and, in some ways, rightly so). Edinburgh do however offer to pay grants in advance into a joint bank

account, operated by the District Council. This saves owners taking out a bridging loan.

If your local District Council does not offer agency agreements, you can ask your local District Councillor to put pressure on the appropriate department to make these agreements, or a similar arrangement, available.

Alterations
Any major alteration to a flat is bound to need permission of some kind or another. You won't need permission to redecorate or replace kitchen units but you will need permission for any work involving rebuilding or demolishing walls, building or moving a bathroom or toilet, changing windows, changing the use of the flat etc. Rebuilding walls will need a **Building Warrant**, changing windows or the use of a flat will need **Planning Permission.** Check also that you are not affecting the **Common Interest.**

Anti-Social Neigbours
Before you embark on any kind of action against your neighbours you should ask yourself *'can the anti-social behaviour be stopped reasonably?'.* For instance, it is not reasonable to demand that a neighbour stops their child crying at night. It is reasonable to ask someone not to start D-I-Y at 11.00p.m. If people can't reasonably

MEET MY SON KNUCKLES!

stop whatever is annoying you then you will need to become less sensitive to the problem.

Action you can take yourself

The first step is to make a polite and friendly request of your neighbour. If you can't trust yourself or your neighbour not to be calm and collected, then write a short letter.

Getting somebody else involved

A local advice agency may be able to offer some help. Failing that a solicitor's letter may be the next step, it having a little more force than your own letter. There may be something in your Title Deeds which stops people doing certain things or putting the property to certain uses. For this, and other cases, you may decide to take private legal action.

The behaviour of your neighbour may constitute a 'breach of the peace' and you might decide to contact the police. This, of course, will probably lead to a complete breakdown of relationships between you and your neighbours.

Taking legal action

If your problem neighbour is a tenant, write to the landlord and ask them to take action - a tenant can be evicted for anti-social behaviour. See **Landlord and Tenant.** If the problem is caused by someone using the house for something other than housing then they should have got planning permission.

If someone is '*continuing conduct which occasions serious disturbance or substantial inconvenience to a neighbour or material damage to his property*' then that conduct can be considered a nuisance under the **Public Health Act 1897** (sections 16-37). You must be able to prove that either the nuisance is something which would affect the normal person or that the nuisance is wanton.

For instance, someone doing trumpet practice in the room above your bedroom at midnight is a nuisance. Practising at 10.00 a.m. would not be a nuisance. If, however, you asked the trumpet player to stop but he carried on just to spite you then that would be a nuisance. *(See Nuisances on page 51 and Taking Legal Action in chapter 9).*

Backcourts

The backcourt is generally **common property** and governed by **common interest** - but check your **Title Deeds**. As common property it is up to every owner to contribute to the upkeep of the backcourt. You might want to come to some arrangement with other owners about this.

Your local District Council may have by-laws which cover the maintenance of back courts. If the District Council has no particular local by-law they may be able to use the **Civic Government (Scotland) Act 1982** which has a section covering the maintenance of private open space 'set apart for the use of owners of two or more properties'. Under section 95 of this Act, there is a duty to maintain such private open space to 'prevent danger or a nuisance to the public'. Any owner carrying out this maintenance can recover an equal portion of the cost from each person entitled to use the open space.

More useful may be Section 92 of the Civic Government (Scotland) Act 1982 which gives the District Council power to act to get the backcourt tidied up. Ask the District Council to use this power: they should be interested in improving the amenity of your area. If your backcourt needs to be improved, you may be able to get an ˇ**environmental improvement grant** from the District Council. Generally, these grants are not given for individual backcourts but only for a whole block of backcourts at once. You may also be able to use repair grants to cover repairs to paths, walls and bin stores if you cannot get an environmental improvement grant.

If you do get an environmental grant, think very carefully about redesigning your backcourt.

> *What happens about rubbish collection and bin stores?*
>
> *Do you want individual binstores (locked?) or one big bin?*
>
> *What does the District Council say?*

> *How much gardening are you prepared to do?*
>
> *Do you want to sit out in the garden? - in which case a seat in the sun would be nice.*
>
> *Do you want whirligig washing lines (prone to being stolen) or conventional clothes poles (take up more room)*
>
> *Do you want somewhere for young children to play?*
>
> *Can you keep people away from ground floor windows?*

You will probably need to hire an architect as there will be a big contract to run and you will also want to look at drainage, 'bin carrying distances' and vehicle access if appropriate.

Basements

Basement owners should note that their flat may be below the **Tolerable Standard** if the floor surface is more than 3' below street level and the rooms are less than 7' high or do not meet Tolerable Standard rules governing dampness, lighting, ventilation etc. as specified in the **Housing (Scotland) Act 1987 (S114).**

Many local authorities have already served **Closing Orders** on basement flats which are either only open to the air on one side of the house or which have small windows. It is possible to get round the dampness problem by installing a **damp proof course** going up the walls to street level. If your flat is partly a basement flat, the Closing Order can only apply to rooms used for sleeping.

If the basement has its own street entrance, the owners may not be obliged to pay for repairs to the common passageway and stairs. Check the **Title Deeds.** Do not be surprised though if these are unclear on the matter - very often basement flats were originally part of a larger ground floor flat from which they were split off. A separate Title Deed may have been drawn up which covers the common repairs problem.

Capital Gains Tax

Capital Gains Tax (CGT) is a tax paid on the increased value of anything you own. If you are an owner-occupier with only one house you will be exempt from CGT when you sell your house. You will get exemption on a second house if it has been lived in by a dependant relative. If you have rented out part of your house, sold part of the house or garden or used part of the house for business purposes, you may get exemption on only part of the house. If you are liable to CGT you will pay tax on the difference between the sale value of the asset and the price you first paid for it. Even if you give things away, you will still have to pay CGT based on a realistic sale value.

Closing And Demolition Orders

A Closing Order *(S114 Housing Scotland Act 1987)* is served on a flat which is considerably below the **Tolerable Standard** and where the necessary repairs and improvements to bring the flat above this standard would be too expensive. The Closing Order prevents people occupying the house after 28 days of the notice being served. (You can be fined £200 if you ignore this order).

You will normally find that all flats in a tenement are served with Closing Orders before a demolition order is made. **The Demolition Order** *(S115 Housing (Scotland) Act 1987)* forces evacuation of the building (you are given at least 28 days notice) and the building must be demolished within 3 months. The order will give specific dates. You can attempt to stop these orders. First of all, you can write to the District Council with your proposals for remedial works and ask for a **suspension order**. You should do this within 21 days but the District Council may give you longer. The suspension order will give you a year to put things right. (You cannot take this step if all the flats have been served with closing orders first).

Secondly, you can appeal against the orders *(S129 Housing (Scotland) Act 1987).* You must do this within 21 days of the order being served or 21 days of being refused a suspension order.

Sometimes the District Council can serve a demolition order with shorter periods of notice if the building is dangerous. *(See Dangerous Buildings).*

Commercial Premises

If you have a shop, pub or other commercial premises in your tenement, you may find it especially difficult to get repairs done. The problem can be even worse if the premises are rented. Commercial leases often force the tenant to pay for repairs and the tenant will never see anything of the increase in value of the property. Another problem is that until recently only house owners could get repair grants. This has now been changed so that if a compulsory repairs notice has been served, commercial premises may be able to get a grant. In practice, you may find that you need to get compulsory repairs notices even for minor repairs. If your Title Deeds say that the cost of repairs is divided according to rateable value then shops and pubs will end up paying the largest share of any repair bill.

Any change of use, for example, from house to shop or store, from shop to restaurant or launderette, needs **planning permission.** Shops, pubs, etc. can still be a nuisance even if they do have all the necessary permissions, licences etc.

A commercial tenant or owner should not use the domestic refuse collection - if they do, complain to the cleansing department as they are trying to avoid paying for a commercial collection.

The Common Stair

The stair is **Common Property** and any repairs will need to be paid for by all the owners. Some owners may try to argue that as they never use the top flight of stairs it's not up to them to pay for any repairs to the top flight but this argument does not hold water. It is certainly against the **Common Interest** and there is probably something in your **Title Deeds** which will cover this specifically. If there isn't, the neighbour refusing to pay will need to take a test case to court. If you get a **Compulsory Repairs Notice**, the District Council will definitely split the cost between all owners evenly or on the basis of rateable values. This means that your neighbour will need to pay first and argue later (probably not winning anyway).

Many District Councils have by-laws governing the cleaning, painting and lighting of common stairs. District Councils are also given powers under the *Civic Government Act 1982.* The first section covers lighting of common stairs. It allows the District Council, although it does not force them, to provide lighting to common property and allows the District Council to force owners to provide suitable lighting. The District Council can charge owners for providing this service (and the charge will be according to Rateable Value). A second section covers the cleaning and painting of the stair *(see Stair Cleaning).* This section also makes it an offence to drop litter on common property (with a fine of up to £200 possibly more). A third section covers fire precautions in common stairs and makes it an occupier's duty to keep the stairs free of anything combustible or anything that blocks access or exits from the close.

Compensation

In some cases you may be able to obtain compensation from the District Council. You won't get compensation just because you feel hard done by but you can claim for the following:-

Compulsory or voluntary purchase by the District Council. Claim site value and 'well maintained payment' or market value.

If forced to move house claim home loss payment, disturbance payment and rehousing by the Council or help to buy a new house.

Effects of construction work carried out by Local Authority. Claim temporary housing and sound insulation costs.

Compensation for effect of new

developments (e.g. motorway building). Claim for sound insulation, and compensation for loss in value of property.

If you cannot sell your house because of the District Council's plan for redeveloping the area claim planning blight compensation.

As you can see the effects on your property must be pretty dire before you get any compensation (but you may still have a claim against your District Council for negligence or other reasons and be able to claim damages).

If you are affected by one of the above, the chances are that your neighbours will be as well. Join or form a **residents association** and fight to get the best deal for everyone affected. You may get a lot more as a group.

The compensation can include the cost of hiring a surveyor or solicitor to act on your behalf. Check with the District Council first.

Compulsory Purchase

Compulsory Purchase Powers (CPO powers) are contained in many Acts but you are most likely to find CPO powers used in Housing Action Areas for Demolition and Highways proposals.

The Compulsory Purchase procedure is the same in almost all cases.

Stage 1.

Compulsory Purchase Order (CPO) is served on owners and occupiers. The order is advertised in at least two local papers for two weeks running.

Stage 2.

A period of at least 21 days is allowed for objections. Objections need to be made in writing to the council department named in the order. You may not have time to write out a full objection but you can put in a 'holding objection' and send further details later.

Stage 3.

The council may discuss objections and amend their CPO in light of the objections.

Stage 4.

A Public Inquiry will definitely be held if someone directly affected has not withdrawn their objection. Everyone affected is allowed to attend and 'third party' objectors may also be able to put their objections.

Stage 5.

The Reporter at the Inquiry sends a report to the Secretary of State who then makes a decision. All those involved are notified of the decision which is also advertised in the Press.

Stage 6.

You can now only appeal to the Court of Session if you think that proper procedures have not been followed.

Stage 7.

You can now claim compensation. If you disagree with the amount you are given, you can appeal to the Lands Tribunal. You can claim for loss of amenity as well as loss of property.

You can object to a CPO if:-

You don't agree with the reason for which the CPO is being served. BUT a CPO normally follows the serving of another notice e.g. for a Housing Action Area and if you haven't objected to that, then it will be difficult for you to object to the CPO on these grounds.

You think there is a need to retain the land for its existing use.

You object to the way things are being done.

You think there is no need for the District Council to buy the land as the existing owners can achieve the same end result as the council.

You can propose an alternative site.

You would be happy about the CPO if minor amendments were made.

You object to the classification of your property, for example, as a house below tolerable standard which gets less compensation.

Legal References

Acquisition of Land (Authorisation Procedure) (Scotland) Act 1947. Schedule I.

Further Reading

Compulsory Purchase in Scotland. Law Society of Scotland (1983)

You and Your Rights in Scotland. Readers Digest (1984).

Crime Prevention

Tenement flats can be made into very secure properties. Flats above ground level can normally only be entered by the close door and many flats are overlooked by neighbours. This makes crime prevention a relatively cheap and easy task. The things you can do include fitting:

Door locks

Make sure you have a mortice lock fitted - the latch of these cannot be pushed back. The lock case can also be protected to prevent the drilling out of the lock.

A good door

Some thieves get in by breaking through the door or breaking the door frame. Solid core doors are generally secure but laminated ply doors are better. Hinges can be reinforced with hinge pins which fit into sockets in the door frame.

Spy holes

There are many spy holes commercially available.

Door entry systems

There are many door entry systems available now and they will cost approx £150 per flat. You can install them yourselves if you are reasonably competent joiners and electricians. The advantage of such entry systems is that they prevent unwanted people getting into the close. If a thief is determined to force the door they will have to do this out on the street, where they are more easily seen. These systems also cut down vandalism in the close and are recommended by the Police.

Window locks

Window locks should be fitted on all ground floor windows and windows accessible from flat roofs etc. If you don't need to open a window you can use a block of wood screwed to the frame and the sash. If you want to allow limited opening you can fit blocks on the top sash and screws about 6" above the runner for the bottom sash.

Window Bars

Window bars may be a good idea in some ground floor flats but don't just lock them and throw away the key! You might need to get out in an emergency. Leave the key to the bars hanging up close by the window or use a combination lock and leave a conspicuous note of the number where it can be seen from inside the house only.

Plastic glazing

Plastic glazing may be used where broken windows are the problem. This material is more expensive than glass and can get scratched and melted. Nonetheless, it can be very useful, especially in close doors and windows.

Drain pipes

You can prevent people climbing drain pipes by either filling the back of the pipe with cement so climbers can't get a grip or fitting a circular guard about 10 or 12 feet above the ground. A special non-drying anti-vandal paint is available, it leaves a mess on anyone touching it. It should only be used on surfaces 8' from the ground, because of the risk of touching it accidentally.

You should also make good use of the Police. The Crime Prevention Officer will give free advice and can recommend specific locks and techniques to suit your problem. You should always report anything suspicious - you needn't give your name to the Police and they do not really mind false alarms if they are raised in good faith. The police sponsor Neighbourhood Crime Watch schemes in some areas. The Crime Prevention Officer would be able to give details of these if a Residents Association were interested.

Further reading:

Beating the Burglars. 'Which?' November 1990 pp 606-611

Keeping the Burglars Out. 'Which?' November 1991 pp 606-611.

'Which?' Magazine is published by the Consumers Association and is usually available at your local library.

Dangerous Buildings

If you find that your home (or the next door building) is dangerous things will happen very quickly and you may have little control over events.

Reporting a dangerous building

During the day, try to contact the council's Building Control or Technical Services Department. Outside normal office hours, or where you have difficulty finding the Building Control number ring the Police. They have emergency contact numbers and will call the appropriate people.

Dangerous Buildings Order

1. A **Section 13 Notice** (1) will be served if your building is dangerous. It is sent to owners, not tenants. The Section 13 notice may be accompanied by another notice to all occupiers requiring them to vacate the property. The Section 13 notice states what work has to be done to make the building safe and gives a time limit for the work.

2. On receiving the notice, you should organise a meeting of owners and decide whether you will go ahead with repairs. Building Control will probably be willing to give you additional advice on costs, etc. Let the Building Inspector know what you want to do. If you want repairs done, they may give you a longer period to do the work, or arrange the repairs themselves.

3. If nothing is done a Section 13 Order may be served on owners to force them to do work. You will be sent a notice of intention (2) first, to give you time to object in writing to the council. Even if the order is made, you can still appeal to the Sheriff Court (3). The order may give you a choice between repair and demolition or you may simply be told that the building must come down.

Evacuating the Building

If you are required to leave the building you will be sent a notice to remove. If you do not move, the council may get a decree for possession from the Sheriff Court and you will be sent a Summons. People evacuated from dangerous buildings should get rehoused (4).

Neighbouring Owners

Some of the notices described above may be served on you, even though your house is not dangerous itself. If the next door building is to be demolished you may need to get your building shored up - if you have a right of support your neighbours may be obliged to pay for this. Otherwise, you may get help from the council or your insurance company. You should also have a structural survey of your building carried out so that you are in a good position to know if and how the demolition has affected your property. Demolition should be carried out according to Building Control Regulations. If you have problems, contact the Building Control Department. (See Demolition).

Finance

If you decide to have work done to repair the dangerous building, you should get grants and loans from the District Council (see Chapter 7).

Whether you repair or demolish, you may be able to claim on your Building Insurance policy. In theory, you are obliged to pay the cost of demolition. Many District Councils, however, will pay for the demolition if you agree to let them have the cleared site.

If you have a mortgage, you should continue payments but inform your Building Society of the problem and they will probably come to some arrangement with you. If you have a council loan, you may find that the council will be willing to write off your remaining debt.

You will also be entitled to compensation if your house is demolished.

Legal References

(1) S13 Building (Scotland) Act 1959

(2) Building (Procedure) (Scotland) Regulations 1975 reg. 52 (Statutory Instrument No.550, 1975).

(3) S16 Building (Scotland) Act 1959.

(4) S36 Land Compensation (Scotland) Act 1973 as amended by Schedule 2 para 14 Housing (Financial Provisions) (Scotland) Act 1978.

Demolition Work

Demolition work must be carried out safely. The Building Control or Technical Services Department of the District Council should control the work and make sure it is done safely (1).

If you see any unsafe practices or are being caused considerable inconvenience, complain to the District Council. Some of the less obvious things to look out for are:-

Adjoining buildings being given support before demolition work starts.

Buildings secured or watched on a 24 hour basis while demolition is taking place.

No bonfires should be lit if there is a smoke control order operational in the area. Burning bonfires should not be left unattended at the end of the day.

Pavements should be kept clear of debris and dust kept down by regular spraying of water.

No piles of debris left in a dangerous state or allowed to build up by unsupported walls.

The site levelled or fenced at the end of the operation.

Legal References

(1) British Standards Institution Code of Practice CP94 1971 (Demolition) Buildings Operations (Scotland) Regulations 1975.

Empty Flats

An owner of an empty flat may still need to pay poll tax and will find it difficult to insure the property, as normal insurance policies lapse if a house is left empty and unfurnished for more than 30 days (but check your policy for details). The owner will still be legally responsible for any damage caused to neighbours' flats by burst pipes etc. If you are having trouble getting hold of the owner at such times, you can ask the District Council to step in and do the work (1) - **the District Council will charge the owner the cost of the work.** If children or vandals are breaking into the flat you can ask the Police to board the house up (again, they can charge the owner) (2).

Squatting is trespass and a criminal offence (3). Call the Police and ask them to take action under the Trespass Act. The owner can use '*reasonable force*' to eject someone but would run the risk of a counter accusation from the squatters to the effect that deliberate injury had been caused. The safest action is to get at least one of the squatters' names and get a **Court Order** made out for eviction. You can demand legal expenses (and other damages) from the squatters but consider using a solicitor as there can be various delaying tactics and other complications. You may be able to freeze out your squatters as gas and electricity boards only have to supply power to legal occupiers. This does not include squatters unless they have a '*licence to occupy*' (the owner's permission to stay there).

Legal References

(1) S20 (3)(b) Public Health (Scotland) Act 1897.

(2) S61 Civic Government (Scotland) Act 1982

(3) Trespass (Scotland) Act 1865.

Further Reading

Housing Action Notes 'Squatting' SHELTER Scotland August 1976

Feus and Ground Burdens

In Scotland, land was often not sold outright but '**feud**'. The buyer was called the feuar and agreed to pay an annual sum (or **feu duty**) to the *feu superior* in return for which they were given the right to the land forever. 'Feu' is the old French for 'fee'. Feuing land was very popular when tenements were first built as it allowed the builder to pass the cost of the land directly on to the people who bought the tenements. Sometimes the feu is called a **ground burden** or **skat.** The feu superior often put restrictions on the land and these will still apply whether or not you pay feu duty. Nowadays, it is felt that feus are outdated and a bother to collect. You can now **redeem** or pay off your feu. An **allocated feu** is one which is applied to individual flats. Each owner is responsible for their own feu. When the flat is sold, the feu must now be redeemed. (This is part of the *Land Tenure Reform (Scotland) Act 1974).* The redemption is normally done for you by the solicitor who does the conveyancing.

Unallocated Feus

An **unallocated** feu is one which applies to the whole tenement. Although every owner is obliged to pay their share, the feu superior can ask one owner to pay the whole feu. The unlucky owner would then have to collect the shares from the neighbours. If you are paying an unallocated feu, you can get the feu **allocated.** You can still redeem unallocated feus although you have to get all your neighbours to co-operate.

Getting the feu allocated

You will need 2 things:-

(1) A '**notice of allocation of feu duty or ground annual'.** This is a form which you can get from Citizens Advice Bureaux, Sheriff Clerks or the Scottish Home and Health Dept., Room 130A, St Andrews House, Edinburgh EH1 3DE.

(2) **The name of the feu superior** or their agent to whom the feu is paid. The factor is not the agent for the feu superior but the factor should give you the appropriate name and address. Ask to see the **cumulo feu duty receipt.** You can also get the feu superior's name from the **Register of Sasines.**

Now fill in the form. It asks for your share of the feu duty, the address and position of your flat (preferably as given in the Deed of Conditions) and the name and address of the person you pay feu duty to (e.g. the factor). Sign the form and send it to the feu superior. The feu superior can only object to the share you have said you will pay - the actual allocation cannot be stopped. The allocation becomes effective on the first term day (e.g. Whitsunday or Martinmas) 3 months after the date you sent off the notice. (If you sent off the form in April, the next term day would be Whitsunday in May but you would not get the feu allocated until Martinmas in November).

Redeeming the Feu

(NB: This also applies to most other ground burdens).

If you are redeeming an unallocated feu you will need to get all your neighbours to co-operate. If your neighbours won't join in, get your own feu allocated and redeem it.

You will need 3 things:-

(1) **A form** for the voluntary redemption of feu duty available from Citizens Advice Bureaux, Sheriff Clerks and the Scottish Home & Health Department, Room 130A, St Andrew's House, Edinburgh EH1 3DE.

(2) **The feu superiors' names.**

(3) **The 'Multiplier'.** This is the amount you would have to invest in 2 $\frac{1}{2}$% Undated Consolidated Stock to get an annual income of £1. This must be multiplied by the feu duty to get the amount of redemption money to be paid. You need the multiplier for the day which is one month before the redemption date e.g. for Whitsunday, 15th May, you would need the multiplier for the 14th April and for Martinmas (11th November) you would need the multiplier for the 10th October. The multiplier is published in the major Scottish newspapers between the 14th April and 15th May and the 10th October and 11th November.

Now fill in the form. It asks for your name and address, the feu superior's name and address, when you wish to redeem the feu (e.g. 'Martinmas 1990'), whether it is a feu duty or ground burden etc. and the annual amount paid.

You will need to calculate the amount of money to be sent. Firstly, multiply your annual feu by the multiplier (remember that if you pay twice a year the annual amount will be twice the amount on your last factor's bill). Secondly, you will need to calculate your arrears. Feus are almost always paid in arrears so you will need to add one more feu duty payment and any other arrears that you may have run up.

Example
You pay feu duty of £5 twice a year. The annual feu duty is £10. The multiplier is shown as being 8.969. You will need to send £10 x 8.969 = £89.69 plus one last payment of £5 to being you up to date = £94.69 in all. Finally send the form and the cheque for the redemption money and make sure it reaches the feu superior before Martinmas or Whitsunday. You will get a receipt from the feu superior and you should keep this safely, passing it on to the new owner when you sell. If you have any difficulty, ask the Citizens Advice Bureau or local advice centre for help. You are unlikely to need a solicitor.

Fire

Strathclyde Fire Brigade estimate that there are an average of 1,250 fires in tenement buildings every year - and every year some 25 people die in these fires. Apart from the loss of life, numerous people are severely scarred or injured and many houses and treasured possessions totally destroyed.

There are simple things that you can do to prevent fire from starting in your flat such as fitting smoke alarms, keeping a fire blanket in the kitchen, and checking the state of your electric wiring.

Your means of escape in a tenement is the common stair - so don't leave things on the stair which can cause an obstruction or be set on fire by vandals. Make sure that there is one window in your flat which you can use as an escape route to the fire brigade's ladders.

If You Discover a Fire
Close all doors, especially the door to the room the fire is in. Don't be tempted to open the door to see how the fire is doing!

Alert the family and neighbours.

Call the Fire Brigade by dialling 999. Give the address of the house clearly.

Get everybody out of the flat involved and the adjoining flats.

If the fire is in a neighbouring flat:-
Alert the family.

Leave the building (if possible) closing all doors.

If smoke or fire prevents you using the stair, go back to your flat, shut all doors, closing yourself in a room which has an opening window, preferably at the front of the building but definitely as far away from the fire as possible.

If smoke starts to get into the room, block gaps with wet towels, woollens, etc.

When the fire brigade arrives, attract their attention and make sure they know where you are.

If you are worried that there is a serious fire risk in your building, or you require advice on fire precautions, do not hesitate to contact the local fire station. The non-emergency telephone number is in the directory under **Fire** and advice is always free.

Front Gardens

The front garden - if you have one - almost always belongs to the ground floor flat and it will be that owner's responsibility to maintain the garden. Check, however, what your **Title Deeds** say.

If you get an **environmental improvement grant** (See Backcourts) you may use some of the money to provide new fencing at the front.

If the front garden becomes very untidy, you may be able to persuade the Council to act under **Nuisance** and **Public Health Legislation.**

History

It is possible to date your tenement from its architectural style alone as well as from the amenities and room sizes dictated by the local Burgh Police Acts. The materials of construction also give a pointer to the date of the building.

Tracing the History of your own Tenement

A good place to start is your local library. Ask if they have collections of old **Ordnance Survey maps, Post Office Maps** or **Feuing Plans.** *(Feuing Plans can also be seen at Register House in Edinburgh).* These will show the gradual build up of your town, fields becoming streets, streets being built and named and renamed. You can also look at old Post Office Directories - these may give the names of the people living in your tenement.

Look at your own **Title Deeds** too - these will start off by describing the land the building is on and who owns it. Check dates against the map. If the deed was drawn up long after the building first appeared on the map then the building may have belonged to one owner for a while, the deed only being drawn up when the owner decided to sell off flats.

The library may also be able to direct you to books describing the history of your area and local history societies. They may even have a collection of old photographs.

Further Reading
The Tenement: A way of Life. Frank Wordsall (good on architectural styles and by-laws).

Middle Class Housing in Britain. M.A. Simpson and T.A. Lloyd, David & Charles, London (1977). (Contains chapter on the West End of Glasgow 1830-1914 describing feus and Deeds of Condition).

The Corporation of Glasgow as owners of tenements. Arthur Kay, Glasgow (1903).

Housing Action Areas (HAAs.)

If the majority of houses in an area lack **Standard Amenities** or are **Below Tolerable Standard** (BTS) the District Council can declare a **Housing Action Area.** There are 3 types of Housing Action Area:-

1. A Housing Action Area for Demolition (HAAD).
A council may declare one of these if the greater part of the houses in the area are **BTS** and the most effective way of dealing with the area is to demolish it.

The area must be in a pretty bad state to deserve this kind of treatment. If you want to fight the District Council's decision:-

Ask the District Council for a copy of the information on which they have based their decision. If officials won't give this to you, ask your District Councillor to become involved.

Get a second opinion on the state of the buildings from an architect or surveyor.

Ask the District Council to pay for this survey. It could save the District Council money in the end if all owners decide to co-operate with the District Council's plans.

See if you can save part of the area and get the HAAD changed to an HAADI (see below).

Try to enlist the co-operation of a local Housing Association. The Scottish Federation of Housing Associations or Scottish Homes can give you some suitable names. If your house is going to be demolished you will be entitled to Compensation and Rehousing.

If you move too soon, you could easily lose substantial benefits. If you have to move before the Compulsory Purchase Order goes through, ask the District Council to buy your house from you as a voluntary acquisition.

2. Housing Action Areas for Improvement (HAAI)

The District Council can declare an **HAAI** if the majority of houses lack one or more **standard amenity** or do not meet the **Tolerable Standard** and all the houses can be improved and brought into a good state of repair.

The advantages of being declared an HAAI are:-

You get a better deal in grants and loans

The whole area gets improved. The area is given a new lease of life and house prices will probably go up.

If home ownership is getting too much for you, the HAAI could give you a chance to become a Housing Association or District Council tenant and remain in your old home.

The disadvantages of an HAAI are:-

The District Council can force improvement on you by using improvement orders (but this could be an advantage for tenants).

The whole area will be in turmoil for up to 5 years.

Quite often a housing association will get involved in the area, buying houses and improving them. Sometimes, the Housing Association will help you improve your own property and offer temporary rehousing while building work goes on. Sometimes, the District Council will do these things themselves. The biggest problem you could face will be lack of action.

To get the best deal:-

Form a Residents Association.

Have regular business meetings between your committee and the District Council or Housing Association.

Hold public meetings at each stage so that everyone has a chance to find out what is happening.

Get the District Council to open a local office or have a regular surgery and get them to open a 'show house' showing people what can be done.

3. Housing Action Areas for Demolition and Improvement (HAADI)

This is a mixture of the 2 sorts of HAA described above. The District Council will demolish some houses and improve others. This can leave people really confused about what is happening to which houses. If you form a Residents Association, you will find it a good idea to visit every owner and make sure they know exactly what is happening to them. The District Council is obliged by law to follow certain procedures and these can take months, more often years to go through *(see page 48 for details).*

The District Council does not have to demolish houses immediately if they are needed to provide accommodation (6).

The District Council can issue **Control of Occupation Notices** (7) to stop new people moving into houses where the District Council has already rehoused one family. This notice is designed to stop people jumping the rehousing queue.

Legal References

(1) S89 Housing (Scotland) Act 1987
(2) S98 Housing (Scotland) Act 1987
(3) S95 Housing (Scotland) Act 1987
(4) S90 Housing (Scotland) Act 1987
(5) S91 Housing (Scotland) Act 1987
(6) S96 Housing (Scotland) Act 1987
(7) S97 Housing (Scotland) Act 1987

Housing Action Area Procedures

All references are to Housing(Scotland) Act 1987

STEP 1

The District Council carries out a survey of all houses to see how many are BTS. A structural survey may be done as well. The District Council will produce a map and schedule showing who owns the houses etc.

STEP 2

The District Council will produce a draft resolution which is sent to the Secretary of State for Scotland. The Secretary of State can decide to modify, accept or reject the resolution but is really only concerned that the District Council has enough money to pay for the grants and loans.

STEP 3

If the Secretary of State approves (or does not comment on the proposal within 28 days of STEP 2) the District Council must then publicise the draft resolution by advertising in two local papers and writing to every owner, lessee or occupier.

STEP 4

There then follows a 2 month period of public consultation. The District Council should hold meetings to tell people what is happening. If you want to object to the HAA Notice you should write to the District Council at this point.

STEP 5

After the 2 month consultation procedure the District Council declares the **Final Resolution** which should take into account only representations made. Again, the resolution must be publicised as in STEP 3. From this point, you are now a **Housing Action Area.** You can start applying for grants and loans and the District Council and Housing Association can start voluntary acquisition. If the District Council wants to use Compulsory Purchase Powers it must follow further steps.

STEP 6

Within six months of STEP 5 (seven months in the case of an HAADI) the District Council must send **Compulsory Purchase Notices** to the appropriate owner *(see page for further details of Compulsory Repairs).*

Insurance

Types of Policy

Most flat owners will have two or three insurance policies covering their flat.

Contents insurance on furniture, carpets, pots and pans etc. (We do not go into detail on these policies as there are many other publications available - your local library will be able to help and you could check back issues of 'Which' also from your local library).

Buildings insurance covers walls, roof,

windows and everything else that is left behind when you move (except aerials and fitted carpets which come under contents insurance). You will be covered for loss or damage caused by storm, flood, fire, water leaks, subsidence, theft, riot and malicious damage. Most policies cover additional risks as well. You will need to check the small print of the policy for these additions.

Some policies are **all risks** but there is normally a list of things not covered. You are generally also covered against **occupiers liabilities.**

Common insurance. This is a type of building insurance which is special to tenements. Most Title Deeds demand that this insurance is kept up but many policies have declined in value to the point where they are virtually useless. These policies generally only give cover against fire (perhaps storm as well). For every £1 you pay in premium you might get less than £1000 of cover. If your premium is £24 each year you will have approximately £24,000 cover. This is very likely leave your property under-insured.

Many people feel that they are over-insured as they have both buildings and common insurance policies. This is only likely to occur if you have a good common insurance policy. If you do, you may be able to negotiate with your insurance company (or building society) to get your individual building policy reduced. Your individual policy will probably still be needed to cover you against additional risks and to compensate for any remaining under-insurance.

It is wisest to have a good common insurance policy as you will then know that everyone else in the block is properly insured. Your individual building policy will not cover you against a neighbour being unable to pay their share of costs because they are under-insured.

Some insurance companies offer a block insurance policy which covers the whole of the tenement in one policy. These policies tend to be cheaper, simply because they are less work for insurance companies to administer. You may need to nominate one owner as 'the insured'. That owner will need to make sure that everyone pays up their share of the policy each year. If one owner drops out, you will need to go back to having individual policies. One major problem

with this kind of insurance is that, as all your existing policies expire at different times, you may have difficulty with timing the start of the block policy. You may be able to get a refund on policies cancelled part way through the year.

What should the policy cost?
A standard buildings policy costs from £2.20 per £1000 insured. A policy with extras will cost from £2.70 per £1000. You need to insure your building for its re-instatement value. This is the cost of rebuilding it exactly as it stands in the same materials. Many people feel this is unfair as if the house burnt down completely, they would be happy to rebuild in cheaper materials - bricks rather than stone. (This is known as the **replacement value**). The problem is that when the building is only partly destroyed, repairs will need to be carried out in the original material and matching the rest of the building. Most tenements are stone and rarely have to be totally rebuilt. If you are insured for the market value of the flat you will probably be drastically **under-insured.**

Many insurance companies will carry out a free valuation survey for you, (they will not be legally bound by it, however). For mid 1991, the cost of rebuilding each square foot of floor area of a stone built tenement flat was estimated to be £85-£100. You will need to measure your flat, including the width of the walls (measure outside and through the close at the ground floor). If your flat is very ornate, or a listed building, or has a lot of expensive fittings (e.g. kitchen, central heating etc.) you may need to add more to the value over and above these costs. Do not forget that the rebuilding cost will increase every year, probably at a higher rate than inflation generally.

Index Linking. If your policy is index linked, you will be covered for any inflation in building costs during the year. If the policy doesn't include index linking, you will need to add about 15% to the reinstatement value to cover inflation. The premium for an index linked policy may be a little higher but you pay it on a slightly lower value so it will probably even out at the end of the year.

New for Old. Some companies offer indemnity insurance. This takes into account 'wear and tear' and 'betterment'. Suppose your decorations are 5 years old when they are damaged. The insurance company may say that your decorations

would have lasted 10 years so you've had half the use of them. Under indemnity cover they would give you half the cost of redecorating. Under **new for old** cover you would probably get the whole cost paid. With building insurance, new for old will only make a difference to claims for redecoration, new bathroom and kitchen fittings, etc. You may feel that it's not worth the extra premium.

Shopping Around. Many insurance companies charge the similar premiums for standard cover. Differences can lie in the risks covered, the extras such as **index linking, new for old** and the **excess** allowed. (Most companies say that you must pay the first £5 and they will pay the excess.) You should check the small print of policies carefully and decide which features you are prepared to pay extra for. 'WHICH' has regular reports on building insurance costs.

Making a Claim on Buildings
Your policy will probably say that you must claim within 30 days of the event. It may take all of that 30 days and more to prepare your claim so start straight away. If your insurance is paid through an agent, such as a building society, you should ring and ask them to send a claim form, otherwise contact the insurance company direct.

If you need to carry out a **holding operation** to stop the problem getting worse don't delay. It shouldn't affect your claim but you could ring the Claims Department of the company and check with them before committing yourself. If the company appears to agree reluctantly that something is a holding operation it would be worth writing to confirm what was agreed on the phone. Whenever you write to your insurance company quote your policy number.

You will next need to get an estimate of the cost of repair. Some insurance companies want two estimates. Send these with the claim form. If you are claiming on an individual policy for a common repair, you may need to get details of the other neighbours insurance policies and the share they are due to pay. If all this seems to be taking more than 30 days, send the claim form with the details of the damage and say you will send other details and estimates later. The insurance company may want to send out a loss adjuster or claims inspector to check the cause and extent of damage. In any event, you shouldn't start work (unless it's an emergency) before the company agrees to pay the claim.

When the work is finished, send a copy of the bill to the insurance company. They will send the money to you. Tell the tradesman if they are going to have to wait for the insurance company to pay up before they get paid. You can use a different tradesman from the one that gave the estimate to do the work. If this tradesman is more expensive than the first, you should check with the insurance company before going ahead. If the second tradesman is cheaper you won't make a profit on the deal   because most insurance companies will still only pay the actual cost as shown on the bill.

The average claim will probably take about six weeks to come through. If you're dissatisfied with the way your claim is being handled, you should first complain to the **Area Claims Manager** at the company. If this doesn't get you anywhere, you could ask the help of your building society or agent, (they shouldn't charge - it's what they get commission for.) The next step would be to write to the **Chief General Manager** at the company's Head Office. After that you are left with the **Insurance Ombudsman** but not all companies are members of the Ombudsman Scheme. The company may have its own arbitration procedure or you could contact the **British Insurance Association, 19 St Vincent Street, Glasgow G1 2DT.** (See also legal advice).

Typical Insurance Problems

Perhaps the most common problem is **under-insurance**. If you are not insured for the full reinstatement value you will be under-insured. Some companies say that this totally invalidates your policy. Other companies will 'average' your claim. They calculate the average by dividing the actual sum insured by the full reinstatement value and multiplying the claim by the amount.

For example, if your reinstatement value is £100,000 but you are only insured for £75,000, you will only get 75% of any claim. If you are claiming £5000 you will only get £3750.

Another problem is that many people think their insurance covers them for all repairs. It doesn't. You are only insured against the specified risks mentioned in

your policy. Gradual deterioration or lack of maintenance cannot be insured against. Dry rot and other timber infestations are almost always considered *'gradual deterioration'*. If your building is in a poor state of repair, this may invalidate your whole policy even if it is not the direct cause of the problem.

Your policy may also be suspended when major repairs are being carried out. If you are planning repairs, write and tell your insurance company. Your policy will also be invalidated if you leave the house empty or leave only a few sticks of furniture for more than 30 days. Your insurance company may still be able to insure you but you could pay a lot more. If you are in doubt, check with your insurance company. When you are buying a house, you become responsible for its insurance from the moment the offer is accepted. You should still maintain insurance on your own home until it actually changes hands in case the buyer backs out of a deal.

Main Door Flats

A main door flat will often have a separate street number from the rest of the tenement but will still be obliged to pay for common repairs. Check your **Title Deeds**, however, as they may not be obliged to pay for the upkeep of the common stair. They may also have sole rights to the front garden and, perhaps, to part of the backcourt.

Mortgage Surveys

There are three basic types of inspection, each having a distinct purpose:

The Mortgage Valuation Report.

The valuation for mortgage purposes is a limited report made for Building Societies, banks and lenders before a loan is made on a property. The inspection will be carried out by a valuer who is usually a qualified surveyor, but it is not a detailed inspection of the property. However obvious defects which are easily accessible (not the loft or under floor areas) should be inspected (if a serious fault in the property is not picked up you may be able to get some compensation).

The surveyor may recommend that a part of the mortgage be retained by the lenders until such time as the work is carried out, and may place a time limit for the work to be completed. You might

not even see the valuation survey anyway as it goes directly to the Building Society or other lender. It won't do any harm to ask for a copy, but don't rely on it as a report on the condition of the building.

Surveyors call this a scheme 1 report and will charge about £70 (plus VAT) for a property worth £40,000.

The House or Flat Buyers Report and Valuation.

This is instructed by the house buyer and provides a direct link with the surveyor. The RICS has produced a standard report form which can be used for flats. The scope of the inspection is quite detailed. You can expect a report on the roof, gutters and downpipes, damp proof course solum space and the common loft space. Internally, woodworm, rot and finishes will be inspected where accessible (floor coverings are not lifted), and wiring,drainage and central heating will also be examined, but not tested. It is usually enough to point out any major defects and is recommended if you are buying a house.

Surveyors refer to this to as a scheme 2 report and will charge about £160 (plus VAT) for a property worth £40,000. In both schemes the charge increases with the value of the property.

A full structural survey.

This is a very detailed report and will take several hours. You will need to decide on the scope of the survey first as other specialists may need to be brought in. The surveyor will normally arrange the opening up of floorboards and hidden areas. Such work can damage decoration and fabric. Services such as electrics, drainage, gas and boilers may also require special testing. You will get most out of one of these surveys if you do your own survey first and then ask the surveyor to look at the elements which most concern you (and perhaps leave other items - such as internal decoration and fittings out of the survey altogether).

This kind of survey is usually charged by the hour and can cost as much as £1000, depending on the level of inspection required.

If things go wrong, you may also take action on grounds of negligence (see Taking Legal Action chapter 9).

We heard of one case recently where someone who had bought a house with a valuation survey found extensive rot caused by rain penetration. Their valuation survey had not picked up evidence of rain penetration but they managed to find out that an unsuccessful buyer's valuation survey had. They were able to use this as evidence that their own

SHADDUP!

surveyor had not exercised 'reasonable care'.

If you do take this kind of action, you may only get damages of the difference between what the house was actually worth and what the surveyor told you it was worth. You will not be able to get damages for the full cost of the repairs.

References:

House Valuations and Surveys - The Choice Available, available from The RICS in Scotland, 9 Manor Place, Edinburgh EH3 7DN Tel:031-225 7078.

Noise

We deal here with noise caused by neighbours, an all too common situation in tenements. Noise from industry, motorways, airports etc. is governed by different laws.

Noise can be dealt with as a Nuisance (1) (see Anti Social Neighbours, and Nuisance and Public Health). You will need to show that the noise causes a disturbance or material inconvenience (e.g. regular loss of sleep). For occasional disturbances you can call the police. They won't say you called them if you ask for your name to be kept out of things but you could still increase the bad feeling in your tenement.

If you complain to the District Council, they may take action under Public Health Acts (1), The Noise Abatement Act (2) or the Control of Pollution Act (3). You can take private legal action yourself under

the Control of Pollution Act (4).

Sound proofing can be very expensive but there are some measures that you can take yourself. Sound travels through air so block up all gaps around pipes with plaster or similar material. Close windows and doors. Vibrations travel through floors and walls and TVs and radios can cause vibration. It is the bass vibrations which travel best. Move your TV, radio or record player away from the wall and don't put speakers on the floor. Soft furnishings, carpets underfelt, curtains, cushions and even bodies absorb noise. A floor to ceiling built in wardrobe filled with clothes could also be a good noise barrier. The main problem with these sound proofing ideas is that they are most effective nearest the source of the sound and so may be of little use to the sufferer. (See Soundproofing Floors on Page 87)

Legal References

(1) S16-31 Public Health (Scotland) Act 1879 (This covers nuisance generally).

(2) S1(2)(a) Noise Abatement Act 1960 (amended so that only 1 occupier need complain for the council to take action).

(3) S58 Control of Pollution Act 1974

(4) S59 Control of Pollution Act 1974

Further Reading

You and Your Rights: An A-Z Guide to the Law. London: Readers Digest Association.

F Lyall (1981) Air Noise, Water and Waste. Glasgow: Planning Exchange.

Nuisances and Public Health

A nuisance can be many things. For conduct which occasions serious disturbance or substantial inconvenience to a neighbour or material damage to his property. *(see Noise and Anti-Social Neighbours).* Other examples of nuisance are:-

> *Careless or faulty use of the property causing real or actual damage to another's property (e.g. overcrowding or a flooding washing machine).*

> *Something that is injurious or dangerous to health (e.g. a leaking toilet).*

A District Council can act on 'reasonable apprehension' of damage being caused. If you are suffering inconvenience rather than actual damage you must prove that the action is wanton, that is, deliberately done knowing that it causes annoyance.

A nuisance must also be more than is reasonably tolerable by a normal person. The law is not there to protect the over-sensitive. If there is a proven nuisance, and the owner does nothing, the District Council can step in and do the necessary work and charge the owner. A tenant can carry out the work and deduct the cost from the rent. An owner can also be fined for refusing to take action to remove a nuisance.

Taking Action

1. Write to the Environmental Health Officer (EHO) with your complaint, keeping a copy for yourself.

2. In the letter, give all details of the complaint:

When and where it is occurring.

How it affects you and other people.

The name and address of the person responsible for the nuisance.

Your name and address.

Times and dates when the EHO can call and see the problem.

3. If you telephone, call in person or someone calls on you note:-

The time and date of the call.

The name and position of the person you talked to. What was said.

4. The best time to contact the EHO is before 10.00a.m. or after 4.00; at other times they should be out of the office following up complaints.

5. If nothing happens as a result of the EHO taking action, get back to them. They often rely on neighbours telling them that the owner has done nothing before taking further action.

6. If the EHO does nothing, contact the Town Clerk or your District Councillor. Give them details of what has happened so far. Consider taking action yourself. You can either take private legal action yourself or you can take action under Section 146 of the Public Health Act.

To do this you will need to get the signature of 10 ratepayers (they need not live in the area or be affected by the nuisance).

7. You may also benefit from the help of a solicitor. (see Taking Legal Action).

Legal References

The Public Health (Scotland) Act 1897

S16 - Defines nuisances

S17 - Compels councils to inspect their area for nuisances and to take action to remove them.

S18 - Gives council power of entry to property if nuisance suspected.

S19 - Says any person can report a nuisance - they need not be directly affected.

S20 - Gives council the power to issue notices etc. requiring nuisances to be removed.

S26 - Local authority can do work if owner defaults

S40 - House in filthy state to be purified if causing ill health.

S146 - Gives ratepayers and others the power to force the District Council to act.

S150 - Allows council to recover cost of doing work - if tenant charged, tenant can recover cost through withholding rent.

Occupiers Liability

The **Occupiers Liability Act** says that you must take reasonable care to prevent injury or damage to people arising from the condition of your property. The law refers to dangers caused by the state of the premises or to anything done or omitted to be done on the premises.

If repairs are the responsibility of the landlord, then the landlord can be held responsible even if they are not 'occupiers' of the property. The law covers any structure, including vehicles, and any property left on the premises. The law would also extend to making a tradesman or architect responsible if they have negligently left premises in an unsafe condition.

An important point to note is that you are also obliged to take reasonable care towards trespassers. In practice, the trespasser would be held partly responsible too.

There has been a court case where a firm which owned a derelict building was prosecuted for injuries caused to an 11 year old boy who had been using the building as a dangerous adventure playground. The amount of compensation was reduced because the boy knew he shouldn't have been there.

Legal Reference

(1) Occupiers Liability (Scotland) Act 1960.

Ombudsmen

There are four Ombudsmen:-

The Parliamentary Ombudsman (officially the Parliamentary Commissioner for Administration) investigates complaints about government departments.

The Local Ombudsman (officially Complaints Commissioner for Local Administration in Scotland) investigates complaints about local councils.

The Insurance Ombudsman employed by a number of major insurance firms - 31 Southampton Rd, London, WC1D 5HJ, Tel:071 242 8613.

The Health Service Ombudsman not dealt with here - further information from Church House, Great Smith Street, London SW1P 3BW. Tel. 071 212 7676.

All Ombudsmen investigate complaints of maladministration in their appropriate fields. Maladministration may arise from such things as unfair discrimination, incompetence, muddle, delay, and failure to follow the authority's own policy or procedures. There could be other causes of maladministration but in any case it refers to the way a decision is taken, not to the merits of a decision. The Parliamentary Ombudsman investigates complaints that injustice has been caused to an individual by the way a central government department have handled a matter. The Local Ombudsman has a similar remit but looks at the actions (or inactions) of your local council. Examples of maladministration include:

Failure to reply adequately and promptly to letters.

General inattention or slowness.

Giving misleading or inaccurate advice.

Rudeness, bias, discrimination or inconsistency.

Failure to have, or to follow properly, reasonable administrative rules and procedures.

These are some of the main things the Local Ombudsman cannot investigate:-

A complaint about something that happened before 16 May 1975.

A complaint about something you knew of more than 12 months before you told the councillor about it. (However, the Commissioner has power to accept your complaint outside this time limit if he considers that there are special circumstances which make it proper to do so).

A complaint about which you have already gone to court or appealed to a tribunal or a Government Minister.

A complaint affecting all or most of the inhabitants of the council's area (for example, a complaint against the general level of rates).

If you are in doubt, send in your complaint.

How to Use the Ombudsman

You must first take up your complaint with the Department concerned. If you get no satisfactory response then you can complain to the Ombudsman but you must first get your complaint countersigned by your M.P. (for a complaint against the government) or a councillor (for a complaint against the council). Complaint forms are available from the Citizens Advice Bureaux or from the Ombudsman. On receiving your complaint, the Ombudsman will decide whether it is a reasonable one for them to take up. You will be informed if they decide to investigate.

One of the Commissioner's staff will then usually arrange to call on you to talk about your complaint. They will also examine the files of the authority and interview people who have dealt with the matter about which you are complaining. All investigations are conducted in private. The investigation will probably take some months. At the end the Commissioner will issue a report which will not usually mention peoples' names. A copy will be sent to you.

The report will say whether the Commissioner finds that injustice has been caused to you through maladministration by the authority. The Local Ombudsman's report is made available to the public for 3 weeks and it will be advertised in the local press. In both cases, it is up to the relevant department to decide what it will do. The Ombudsman cannot force them to do anything but you may get compensation, or an apology, or you might actually get things put right.

Where To Find Them

The Parliamentary Ombudsman, Church House, Great Smith Street, London SW1P 3BW. Tel. 071 212 7676

The Commissioner for Local Administration in Scotland, Shandwick Place, Edinburgh EH2 4RG. Tel 031 229 4472

Pests

In a tenement, an infestation of vermin or insects in one flat can quickly spread to another. If you know where the insects or mice are coming from, ask the **Environmental Health Officer** (EHO) to take action on a **Nuisance.**

The EHO will be able to give you advice on how to treat pests in your own home. Sometimes they will deal with the problem for you but they may make a charge. Private firms can also get rid of pests at a price.

Pets

There are various things that can be done with anti-social pets:-

Dangerous Animals. Anyone keeping a *'dangerous wild animal'* should have a licence. Check with the police (1). A dangerous dog should be reported to the police. The Police can act against owners of animals which are dangerous, a nuisance or not kept under control or which give reasonable cause for alarm or annoyance (3).

Animals that mess the stair or bark constantly or cause any *'annoyance in the home'*. Contact the police first but you may need to take private action yourself under the *Civic Government Scotland Act 1982* (4). The Environmental Health Department will act against owners who allow their dogs to foul footpaths, grass verges, pedestrian precincts or children play areas (5).

Your deed of conditions may prohibit the keeping of some or all types of animal.

Legal References

(1) Dangerous Wild Animals Act 1976

(2) S(2) Dogs Act 1871

(3) S49(1) Civic Government (Scotland) Act 1982

(4) S49(2) Civic Government (Scotland) Act 1982

(5) S48 Civic Government (Scotland) Act 1982

Planning Permission

You are obliged to get planning permission for any change to the outside of the building and for any **change of use** (1). People who own detached, semi-detached or terraced houses are allowed to do certain things without first asking for planning permission. This is called **permitted development** and allows the building of house extensions, garages, etc (2). Permitted development does not generally apply to flats, but stonecleaning is a permitted development unless your property is in a conservation area or your building is listed. Satellite TV dishes are often considered permitted development but the regulations are complex so always check first with your planning department before you fix anything to the outside of your building.

Tenement flat owners must get planning permission for any alteration to the outside of the building. This includes:-

Window replacement.

Reroofing.

Adding close doors.

Backcourt improvements.

Demolishing chimney heads.

You will need to pay for the processing of your planning application (even if it's refused) (3). The fee for alterations to an existing single house is £46. If you are making an application for your close, then the charge will be £92. (This fee is subject to change). Change of Use applications will also cost £92. Glasgow District Council has a very good policy for flat owners - they will allow you automatic free planning permission for window replacement providing the new windows conform to a specified style.

Making a Planning Application

To make a Planning Application, you will need to fill in a form obtained from the Planning Department. The Planners will help you with any queries about the form. The form will need to be accompanied by plans showing what you intend to do and where you intend to do it. You will normally have to show your property marked out on a large scale Ordnance Survey Map.

The applicant is responsible for neighbour notification (4). Generally, you must send neighbours a letter saying merely that you are making a planning application and describe the proposal allowing a minimum of 14 days for

comments or objections to be made. Roads, Building Control, Environmental Health, the Fire Brigade and any other interested party may also be asked to comment.

The planner may suggest that changes are made to the proposal in order to take into account peoples' comments.

Unless the application is withdrawn, it will need to go to the Planning Committee to be approved or rejected. Both the applicant and objectors can ask to be heard at the Committee.

If you don't like the decision, you can appeal to the Secretary of State. If you need help at any stage, you may get help from Planning Aid.

Objecting to a Planning Application

Anyone can object to a planning application but only neighbours will be formally notified by letter. The letter will tell you where you can see a copy of the application. Write to the Planning Department if you want to object. If changes are made, you may not hear about the changes unless you specifically ask the Planners to tell you.

Listed or Historic Buildings (5)

If your building is listed that is, it has been designated a building of historic or architectural interest - you will find planners much stricter when assessing your proposals. You will also need to get listed building consent. This may even stop you painting your front door a certain colour and stop you making internal alterations. If you own a listed building you may be able to get extra grant money from Historic Scotland. The Planning Department will be able to help you with this, and may give you money themselves.

Conservation Areas (6)

The area you live in may be a designated conservation area. In these areas, planners often have additional powers to maintain (and improve) the amenity of the area. You may find yourself restricted in the colours you can paint your building and trees will be protected from being cut down or pollarded etc. If you have any doubts about needing planning permission, you should call at the Planning Department and ask their advice. Planners will be very willing to discuss things with you and will tell you what you need to do to ensure that your planning application goes through as speedily as possible.

Getting planning approval does not exempt you from needing building warrants; neither does it automatically qualify you for getting a grant.

Legal References

(1) S20(1) Town & Country Planning (Scotland) Act 1972

(2) Schedule I, Class I General Development (Scotland) Order 1981 amended 1983 and The Town & Country Planning (General Permitted Development)(Scotland) Order 1992.

(3) Town & Country Planning (Fees for Applications and Deemed Applications) (Scotland) Regulations 1983.

(4) Town & Country Planning (General Development Procedure)(Scotland) Order 1992

(5) S52-56 Town & Country Planning (Scotland) Act 1972.

(6) 2(1) Town and Country Amenities Act 1974

Rehousing

The District Council is obliged to offer you rehousing in the following circumstances:-

> *Your house has been compulsorarily purchased or demolished*
>
> *In this case the District Council must offer you 'suitable alternative accommodation' within 'reasonable distance' of your existing home 'as far as practicable'.*
>
> *You become homeless and you have children or you are elderly, pregnant or otherwise 'at risk' and the District Council does not believe you to be 'intentionally' homeless.*

The District Council may (but is not obliged) to offer you rehousing in other circumstances. The District Council cannot refuse you rehousing simply because you are an owner-occupier. Many District Councils offer extra priority to people who have been on the waiting list for some time so get your name on the waiting list as soon as you see any need for it.

Many District Councils offer temporary rehousing or decanting for owners who need to leave their houses while building work is carried out. If the District Council will give you temporary rehousing, you may still not get a first class offer. The District Council will

normally do its best to give you something suitable but some District Councils prefer to let the good houses to people who are going to live in them for many years. In these cases, owner occupiers can get a worse deal than tenants. Tenants should expect their landlord to offer them alternative accommodation and should not have to pay any more for it. Tenants need only pay one rent and should not lose security of tenure through temporary rehousing.

If you do move into a District Council house for any length of time you will be obliged to pay rent but you may also be eligible for Housing Benefit if you have a low income.

Legal References

(1) S36 Land Compensation (Scotland) Act 1973

(2) Homeless Persons Act 1977

Rented Property and Repairs

Both landlord and tenant have a responsibility to carry out repairs. In residential property, the landlord is responsible for all repairs to the structure and the outside of the house, for sanitary fittings, gas and water pipes, electrical wiring, fixed fires and water heaters (1).

There are very few exceptions to this rule. The tenant is responsible for repairing any damage caused by them, their household or their visitors. The tenant is also responsible for repairing anything which belongs to them, for example, a cooker, or a portable electric fire. This means that if you are trying to organise common repairs, you will need to get the landlord's approval, but as a matter of courtesy, you should also let the tenant know what is happening.

Although tenants are not responsible for common repairs, they do have certain rights. The landlord must provide the tenants with suitable alternative accommodation if the house has to be vacated. The tenant will only have to pay rent on one house and the amount paid should be the lowest of the two rents. The tenant cannot prevent the landlord carrying out repairs but should be given reasonable notice. If a tenant refuses to allow access to the landlord's workmen, the landlord can get a court order.

Both tenants and neighbours can force a landlord to carry out repairs by using

compulsory repairs notices or **Public Health legislation** *(see Nuisance).* Tenants can also, as a final resort, withhold rent and use the money to pay for repairs.

Tenants have the same responsibilities as owners to be good neighbours - *(see Stair Cleaning, Anti-social Neighbours, Nuisance).* If tenants are anti-social you can complain to the landlord who may be able to evict the tenant if the complaint is serious.

It is very difficult to stop someone from renting out their flat but you may be able to do so if the flat is let as bed sits or is let to more than one household. In this case the house may be defined as being **multi-occupied** and need **Planning Permission** and the approval of the Environmental Health Department. In law, a household is generally defined as people who eat meals together. A student flat might not be multi-occupied according to this definition but if each person has a separate rent book it would be difficult for them to claim they are indeed a single household.

There may be something in your **Deed of Conditions** preventing the renting of flats.

Residents Associations

There is often a time when you will get a much better deal as a group than as an individual. A **residents association** can put a lot more pressure on the Council or other bodies than you can individually. The association may also hold regular meetings where councillors and council officials are invited to come and speak and listen to complaints. Your local Council may also keep the residents association informed of proposals, requests for planning permission and other changes. Of course, you will also be able to get such news and help from your local Community Council but sometimes, the Community Council is not in a position to deal with the type of issues a residents association might take up. This will be the case especially where 2 or 3 streets are affected by something such as Housing Action Area proposals.

If there is no residents association in your area, there is nothing to stop you starting one yourself. The Education Department will probably give you use of a school hall or classroom free. The local advice and information office may be able to help by giving you advice on running meetings, getting grants, etc. The local community centre may also offer facilities and there may be a community worker there who can give you a lot more assistance and practical help. The Social Work Department or Community Education Department may also have a community worker and may also be able to give you a small setting up grant. You could also ask for help from the Housing Department or other Council departments. The support of the local Councillor will be invaluable here. There may also be a local Council of Social Service who will be able to help you.

Further Reading

Leaflets from SCVO and Community Projects Foundation,

How to run a Pressure Group, by Christopher Hall. Dent. 1974.

Right to Light

Unfortunately, there is no specific law in Scotland which says that you have a right to light. There may be a right to light and air, protected by servitude. There may be something in the **Title Deeds** and anyone erecting a new building must get **Planning Permission** first and you can legitimately object on the grounds that you will lose light. This doesn't help you though if your problem comes from overgrown trees *(but these might be a Nuisance for other reasons).*

(See also Planning Permission page 53)

Right of Support

An owner of a tenement flat can claim a **right of support** from flats underneath through **Common Interest, Nuisance** and through **Title Deeds**.

You also have right of support from a next door building. If that building is demolished, the owners must ensure that no damage will occur to your building. If damage does happen, the demolishers could be held to be negligent or to have caused a **nuisance**. There may also be something in the Title Deeds giving you a right of support.

(See also Demolition and Dangerous Buildings).

Stair Cleaning

Everyone is expected to take their turn at cleaning the stairs including the landing they live on and the stairs running down to the next landing.

Some effective tactics for dealing with non-cleaners are:

A reminder card to be passed on from owner to owner in turn.

Offering to clean the stairs for them - at a reasonable charge!

Ask the Environmental Health Department to take action.

The Environmental Health Department may use local by-laws but they also have power to take action under the *Code of Civic Government Act* (1). This says *"it shall be the duty of the occupier to keep the common property clean to the satisfaction of the District Council"*. A fine of up to £50 can be imposed on someone who refuses to clean stairs. The same law can also be used to make people redecorate the close (2) and someone who drops litter in the stair can be fined £200. (3).

Legal References:

(1) Section 92 Code of Civic Government Act 1982.

(2) Section 92 (6) Code of Civic Government Act 1982.

(3) Section 92 (9) Code of Civic Government Act 1982.

Standard Amenities

This is a legal and housing term which is simply translated as 'all mod. cons.' It is defined as having:

i) a fixed bath or shower and an adequate supply of hot and cold water supply at the bath/shower

ii) a wash hand basin (with adequate hot and cold water supply)

iii) a sink (with adequate hot and cold water supply)

iv) a water closet. (1)

You can get grants to install standard amenities if you do not already have them *(see p 31).*

Legal References

(1) Schedule I Housing (Scotland) Act 1974.

Tax Relief on Loans

If you have a mortgage, you should be getting tax relief on the interest on loans up to £30,000. If you are on the basic rate of tax, this is currently worth 25% of the actual amount of interest you pay in a year. You will probably be getting tax relief by the MIRAS scheme (Mortgage Interest Relief at Source). Under this scheme, the Building Society claims your tax relief from the Government directly. You do not see the benefit of the

tax relief in your wage packet but you pay less to the Building Society. The process is automatic and you will be in the MIRAS scheme even if you are not working or paying tax.

Tolerable Standard

This is another legal and housing term. You will often see the abbreviation **BTS** which means '**below the tolerable standard**'. If your house is BTS it may also be defined as '**unfit for human habitation**'.

To be above the tolerable standard (1) means that the house is:

(a) structurally stable

(b) substantially free from rising or penetrating damp

(c) has satisfactory provision for natural and artificial lighting, ventilation and heating

(d) has an adequate piped supply of wholesome water available within the house

(e) has a sink provided with a satisfactory supply of both hot and cold water in the house

(f) has exclusive use of a water closet which is satisfactorily situated within the house

(g) has an effective system for drainage and disposal of foul and surface water

(h) has satisfactory facilities for the cooking of food within the home

(i) has satisfactory access to all external doors and outbuildings (1)

If your house is BTS you should be able to get grants to improve it (see p 31).

The Secretary of State has the power to change the definition of the Tolerable Standard but has not done so (despite pressure from housing groups to have a bath or shower included in the standard).

Legal References

(1) Part 4 Housing (Scotland) Act 1987.

VAT

You will pay VAT on materials and labour costs for carrying out repairs and improvements. Beware of using firms which do not charge VAT. You are only exempt from VAT if you are virtually rebuilding a listed building.

Improvements used to be exempt from VAT but this was changed in the budget of March 1984. You have always been liable for VAT on repairs. It seems ridiculous that the government should tax people who need to carry out essential work especially when the government itself puts so much money into grants.

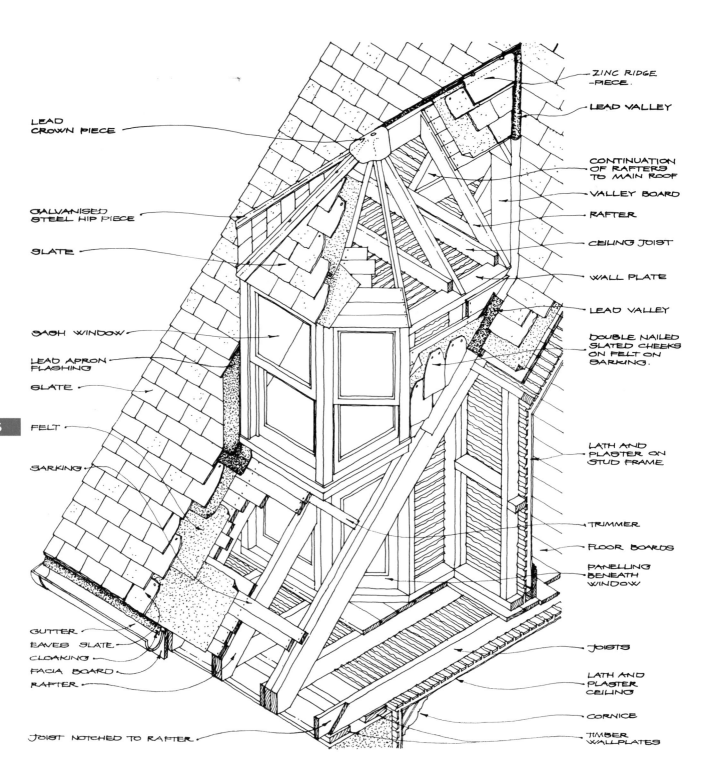

LEAD
CROWN PIECE

GALVANISED
STEEL HIP PIECE

SLATE

SASH WINDOW

LEAD APRON
FLASHING

SLATE

FELT

SARKING

GUTTER
EAVES SLATE
CLOAKING
FACIA BOARD
RAFTER

JOIST NOTCHED TO RAFTER

ZINC RIDGE
PIECE.

LEAD VALLEY

CONTINUATION
OF RAFTERS
TO MAIN ROOF

VALLEY BOARD

RAFTER

CEILING JOIST

WALL PLATE

LEAD VALLEY

DOUBLE NAILED
SLATED CHEEKS
ON FELT ON
SARKING.

LATH AND
PLASTER ON
STUD FRAME

TRIMMER

FLOOR BOARDS

PANELLING
BENEATH
WINDOW

JOISTS

LATH AND
PLASTER
CEILING

CORNICE

TIMBER
WALLPLATES

11.1 DORMER CONSTUCTION

The Roof

The roof is likely to be the single most important (and expensive) item to keep in a good state of repair. A leaking roof will not only cause inconvenience to top floor residents, water penetration of any kind can lead to outbreaks of rot which eventually weaken the whole structure of the tenement.

There are many different elements which make up the roof: the roof and its finishes; the flashings and skews; the chimneys and flues; the loft and roof trusses. This chapter covers all these elements *(gutters and downpipes are described in the Plumbing chapter on page 89)*. All of these elements are important if you are carrying out a major roof repair.

Roof Inspection

It is often very difficult to find out exactly what is wrong with a roof by inspecting it from the ground. Binoculars can reveal blocked or leaking gutters, but can rarely tell you precisely the state of the slaterwork, flashings or chimneyheads. A proper inspection will necessitate going onto the roof itself. If you are thinking of doing this yourself you should do it with a partner and be sure to wear a proper body harness and rope. Rubber soled shoes are also essential. The difficulty with walking on a roof is proportional to the steepness of the roof and the condition of the slates or tiles. Never go up if it is wet. Always make sure both feet are planted on the slates as a slipped or broken slate is quite common. An inspection of the loft space should reveal any signs of rain penetration, and a close inspection of the joist ends should reveal any serious outbreaks of rot.

A Slated Roof

The traditional **Scottish slate** has always been much thicker and smaller than its Welsh counterpart which was used in the rest of the UK. This fact, together with the harsher Scottish climate, has influenced the construction of roofs through the ages. In order to achieve a proper run off, the roof pitch is usually no less than 30 degrees. In addition, because of the smaller slate size, slates are nailed into **sarking**

boards which cover the entire roof. The sarking is made from closely fitting timber planks nailed to the roof timbers. Each slate has at least one fixing hole at the head and is nailed directly into the sarking. Each nail hole is covered by at least two rows of slates *(see fig.11.3)*. Each row of slates is laid in a staggered fashion and tight to one another.

If the pitch of the roof is steep (in a **turret** or **dormer** see 11.1), the slates should be double nailed or side nailed to prevent wind lifting the slate and pushing it over the next one.

The first row of slates is at the **eaves** *(gutter level)*. These will have an initial undercloak course of slates to ensure water does not percolate through the

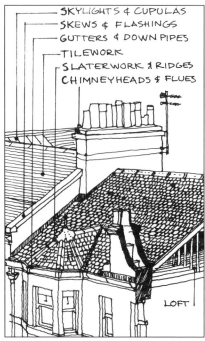

— SKYLIGHTS & CUPULAS
— SKEWS & FLASHINGS
— GUTTERS & DOWN PIPES
— TILEWORK
— SLATERWORK & RIDGES
— CHIMNEYHEADS & FLUES

LOFT

11.2 THE ROOF

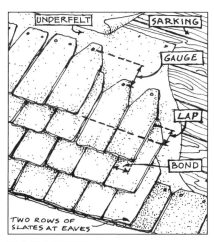

UNDERFELT   SARKING

GAUGE

LAP

BOND

TWO ROWS OF SLATES AT EAVES

11.3 SLATE ROOF

SINGLE SLATE REFITTED BY SLIDING SLATES IN COURSES ABOVE TO THE SIDE. ONLY POSSIBLE IN SINGLE TOP HOLED SLATES

11.4 SLATE REPAIR

joints of the first course.

The **ridge** of the roof is usually made of zinc, or more recently aluminium, and is held in place by metal **ridge clips** nailed through the ridge into the ridge pole *(see fig.11.11)*. The same metal ridge is often used on **hips**, with the joint between the two ridges covered in a **lead crown piece**. Smaller hips and ridges on bay roofs and dormers may be made from leadwork, set under the slaterwork to form a neater finish.

Reslating versus Retiling

Once you or your professionals have carried out a full roof inspection you will need to decide on the best course of action. Minor repairs are to be expected in the life of a slate roof. A good overhaul may be all that is necessary to keep the rain out for another five years. But if it is clear that the roof needs to be renewed then you will need to decide whether to reslate or retile. Although it is cheaper to retile using concrete tiles, slaterwork is not much more expensive. A newly slated roof could be expected to last longer than a concrete tiled roof (after all many have already lasted a hundred years). Furthermore, always check first with the Planning Department, as they may have conditions on the type of roof finish you can use.

Repairing a Slate Roof

Since most Scottish slates are single nailed at the head, it is possible to carry out a repair by sliding the slates to either side and nailing the new slate into place. The slates are then swivelled back into place (see fig.11.4). In many cases, what may seem like one broken slate, can turn into a large area of reslating.

New slates will always be required, so make sure your slater has enough. A small supply in the loft space is useful in emergencies.

Always make sure that slate replacements are the same type and thickness of the existing slates.

If the slater is doing a simple repair, they

are unlikely to erect a scaffold. A temporary roof scaffold can be arranged, or a roof ladder employed. A body harness and rope should be used for additional safety.

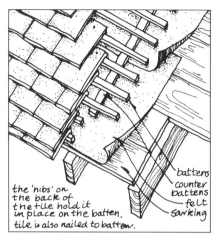

the 'nibs' on the back of the tile hold it in place on the batten. tile is also nailed to batten.

battens
counter battens
felt
sarking

11.5 CLAY PANTILES

Reslating a Roof

A full scaffolding will be required for such extensive works.

Unfortunately, the main Scottish slate quarry at Ballachulish has been shut down but new slate which is suitable for Scottish construction is obtainable from quarries in Cumbria.

Alternatively, slate can be re-used, provided the slates are graded into sizes.

New nail holes may be required as the old slates may need recutting. These regraded slates can be obtained from people who specialise in slate recovery. (It is interesting to note that a lot of Glasgow's slate roofs have been stripped and retiled; the old slates ending up reslating roofs in Edinburgh).

The procedure for reslating are similar to those described under the Retiling a Roof section, otherwise the main steps may be summarised as follows:

Once the old slates have been removed, the timber sarking should be cleared of all old nails and repaired where necessary. In some cases it may be worth replacing this sarking completely to ensure a good fixing for the new slates.

The roof is then felted. This protects the roof to some degree during reslating, and provides a secondary barrier to rain for the completed job.

The flashings are then laid at junctions to skews, chimneys and at valleys.

The slates are laid and nailed into place using the correct lap and bond (see fig.11.3).

Standards

Although there is a British Standard for slaterwork and tiling (BS5534 Pts.1&2), it is always best to try and select a good slater to start with. Good slaters are hard to come by. Many building firms will say they have 'slaters' but because of the relatively small amount of slaterwork done these days they may be very inexperienced. Make sure the following items of specification are clarified:-

What nails will be used? Galvanised nails are frequently used but can start to rust relatively quickly. A better, but more expensive alternative are copper or stainless steel nails.

There should not be a gap between the slates. If there is, the slates can move sideways and rain can drive up between the joints.

It is essential to have the correct lap. This will vary with the slate size, but in all cases the nail hole of each slate is covered by two rows of slates.

A Tiled Roof

Clay roof tiles were used in the 30s and 40s *(often referred to as Rosemary Tiles see 11.4).* **Concrete roof tiles** came into use after 1945 and now hold a virtual monopoly over the roofing industry. The concrete tile now comes in a wide variety of shapes and finishes. (see 11.6)

Recent innovations using resin reinforced tiles have led to very thin **simulated slate tiles.** The tiles are clipped or nailed to timber battens which are placed at sufficient centres to allow the tiles to lap over one another. The tiles are designed to interlock with each other, forming an inbuilt watergate between each tile.

In Scotland, timber **sarking** continues to be used. In order to provide an air gap between the underfelt on the sarking and the tiles, additional **counterbattens** are laid at right angles to the **tile battens**. In roofs built before the 1970s single battens were common.

Repairing a Tiled Roof

It is usually possible to obtain tile replacements. But some makes of tiles have been discontinued and we can expect this trend to continue. Always try to store a good number of spare tiles in

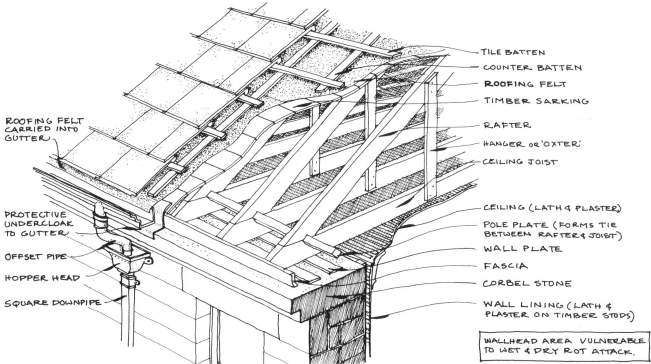

ROOFING FELT CARRIED INTO GUTTER

PROTECTIVE UNDERCLOAK TO GUTTER

OFFSET PIPE

HOPPER HEAD

SQUARE DOWNPIPE

TILE BATTEN
COUNTER BATTEN
ROOFING FELT
TIMBER SARKING
RAFTER
HANGER OR 'OXTER'
CEILING JOIST

CEILING (LATH & PLASTER)
POLE PLATE (FORMS TIE BETWEEN RAFTER & JOIST)
WALL PLATE
FASCIA
CORBEL STONE
WALL LINING (LATH & PLASTER ON TIMBER STODS)

WALLHEAD AREA VULNERABLE TO WET & DRY ROT ATTACK.

11.6 WALLHEAD AT GUTTER

the loft space. Tiles are relatively easily replaced, provided the clips can be removed, or adjacent tiles pushed back.

Retiling a Roof

While tiles are a very cost effective solution to roof renewal, they are not always appropriate because of their large size. Some old tenement roofs can be made up from a lot of smaller roofs, **dormers, bay roofs** and even roofs with a curvature. It may be more appropriate to use slates or small clay or concrete pantiles in these areas. If large tiles are used on small areas, they need to be cut to fit the shape of the roof. The small cut tiles are often difficult to secure properly, and could cause a greater maintenance problem.

Reroofing Procedure

The stages of work for reslating or retiling are similar, the work covers the following areas:

The scaffolding is erected and boarded out. The skip is sited (after approval from the roads department) and the rubbish chute secured. The chute and skip can often be completely joined with a tarpaulin to prevent dust escaping.

After the chimneys are tested, defective chimneys are taken down and put into the skip. They are then rebuilt and new **copings** and pots fixed. If this work is carried out early in the contract, it avoids damage to the new roof work.

The slates are then stripped from the roof. If they are being discarded, they should be placed in the chute and never dropped into the skip from the scaffold.

The sarking is inspected and repaired. Old nails are hammered down. The roof is then felted (and counterbattened if tiling).

Many builders feel that felting is adequate to prevent further rain penetration. However we recommend that even a felted roof should be tarpaulined to provide adequate protection.

Rot repairs are carried out as required, usually at the **rafter feet.** Once all the rot repairs are complete, the last of the battens and timber fillets for flashings are fixed.

Flashings are laid at all roof joints, skews, chimneys, valleys etc (see flashings on page 60).

Gutters are fixed into place and to the correct fall (see gutters on page 92).

What to Look Out For

There is a tendency to condemn outright a tiled roof which is not **counterbattened.** Our view is that if the roof is not leaking it is best left alone. There are a number of other defects which are potentially more serious:-

Some old concrete tiles were **sand faced.** They have a tendency to hold moisture which over the years allows a build up of moss growth. This can increase the risk of frost damage to the tiles.

Tiles may be blown free or loose in high winds. Nowadays all tiles should be **clipped,** or at least nailed. It is not uncommon to find rows of tiles without fixings, this would have been done to facilitate maintenance, but if large numbers of tiles are left unfixed whole sections of the roof could blow off.

Ridge tiles and **verge tiles** are most likely to be lifted in high winds: make sure they are well secured.

Cracked or chipped tiles. Some cracks will be obviously apparent, but the covered area at the interlocking joint between each tile can be cracked and allow water to penetrate.

Sagging or uneven roofs. This can be the result of movement in the roof trusses. Check timbers in loft space and any sign of movement in the stonework at the eaves. Rot at the rafter ends can lead to substantial movement.

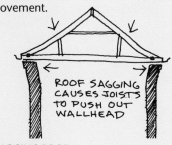

ROOF SAGGING CAUSES JOISTS TO PUSH OUT WALLHEAD

SAGGING ROOF

Lead Flashing Codes		
Approximate Weight Kg^2	Colour	Number
15	Green	3
20	Blue	4
25	Red	5
30	Black	6

The roof is then tiled or slated. **Roof vents, slate pieces** and **skylights** installed.

Ridge tiles or **ridge cappings** are then fitted and the roof cleaned down. Gutters should be tested with water to ensure they are leak free and flowing correctly.

Flashings

A flashing is required at any joint between the tile or slaterwork and a junction with another roof slope, skew, wall, chimneyhead, or pipe. It is one of the most vulnerable areas of the roof. There are small differences between flashings used on a tiled roof and those used on a concrete roof (the drawings below show some typical flashings for a slated roof).

Material for Flashing

The following materials are used in flashing work:

LEAD: this is the commonest material because of its long life. There are different weights of lead sheet, each given a number and a colour code. Code 5 lead (colour coded red at the edges) is in general use for flashings, but heavier lead may be required at parapet valleys and flat roofs (see table). Lead should be laid over an even surface and be secured with copper nails. If the sarking is uneven, it can be levelled with ply sheeting first. The lead should be laid on a separating layer such as building paper or geotextile felt (such as Lotrak non-woven Trevira underlay). To prevent staining of any highly visible leadwork, patination oil is painted over the surface of the lead.

COPPER: this is little used today, but common in the 1950s. Copper corrodes if placed in contact with steel, aluminium or brass.

ZINC: this is most commonly used for preformed **ridge flashings** but has a shorter life than aluminium.

ALUMINIUM: if used as a **ridge flashing**, it should be secured with

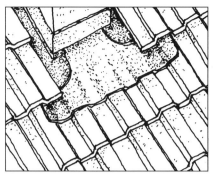

11.7 APRON FLASHING

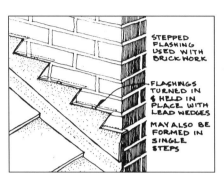

11.8 STEPPED FLASHING

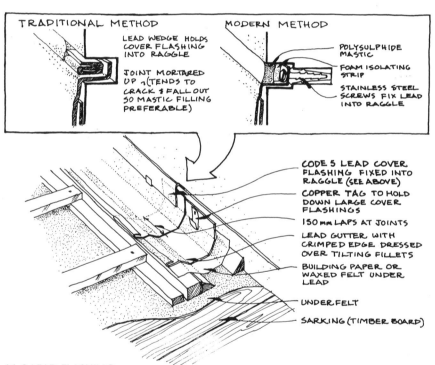

11.9 LEAD FLASHING

aluminium clips and nails to prevent bimetallic corrosion. Aluminium can also be used as a valley or skew flashing (trade name Alcanflash), and also comes in very thin sheet, bonded with a bituminous material, often used on patch repairs.

NURALITE: the trade name for a tough bituminous based flashing material. It is formed to shape by gently heating it, otherwise it can be used in much the same way as lead. Often used where lead theft is a problem. It has a much shorter life than lead.

PLASTIQUE: a bituminous flexible material, most commonly used for short term repairs as it can be heat welted to shape and bonded over old lead flashings.

Chimneyhead and Wallhead Flashings

The lead flashing forms a **secret valley** where the roof slope meets the stonework. To prevent water penetrating under the tilework, the lead is formed over a **fillet** (called doubling in slaterwork), then turned over to form an additional **watergate**. To allow the lead to move and to simplify fixing, a lead cover flashing is **raggled** into a straight groove set in the wallhead, and securely fixed (see fig.11.9). However, many new chimneyheads are built of facing brick and in this case the cover flashings should be **stepped** (see fig.11.8). At the base of a chimneyhead the flashing that overlaps the tiles is called an **apron flashing.** The apron is laid over ply packing (see fig.11.7).

Ridges

Terracotta or concrete ridges come in a variety of shapes. They are more frequently found on tiled roofs. They are fixed in place by being bedded in cement applied to the top row of tiles. When used on hips, the bottom ridge tile is held in place by a hip iron. Because of the problem of ridge tiles lifting in high winds, a dry ridge system was

60

developed. This fixes the tiles without the need for cement.

The majority of slated roofs are capped with preformed zinc or aluminium flashings, held in place by galvanised (or aluminium) ridge straps nailed through the capping into a timber ridge pole *(see fig. 11.11)*.

Lead is used for better class work. At hips the lead can be made into a secret flashing in two ways:

By using a bottle flashing with watergates on either side *(see fig. 11.12)*.

By using individual lead soakers, set under each course of slating and hidden from view.

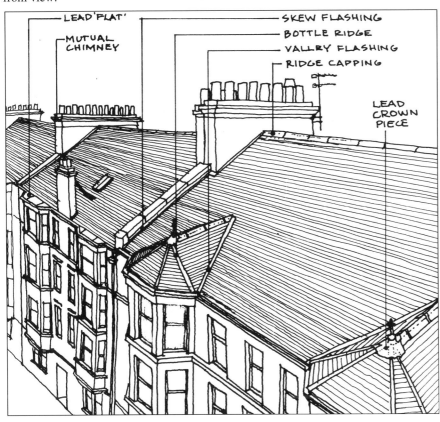

11.10 SLATE ROOF FLASHINGS AND VALLEYS

11.13 LEADWORK AT SKEW

11.14 CEMENT FILLET

61

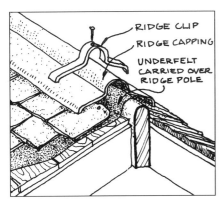

11.11 RIDGE FLASHING

11.12 BOTTLE RIDGE

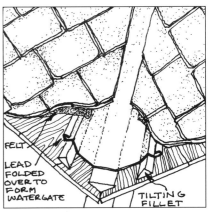

11.15 VALLEY FLASHING

Skews and Valley Flashings

The **skew** forms a valuable **fire stop** as well as a physical break between tenement properties. The capping is usually made from stone, with a special **clubskew** at the base to prevent the stone cappings slipping down.

In good construction, each **capping stone** is interlocked with the next one to provide a raincheck. The lead flashing at the skew is formed in the same way as a wallhead flashing, but cover flashings are sometimes omitted because the skew stone does not allow enough space for fixing *(see fig.11.13)*.

On the East coast of Scotland, flashings may be formed by turning the slaterwork up at the skew, and forming a cement fillet joint. This can be quite effective provided it is done correctly *(see fig.11.14)*.

Because the skew has often been a point of water ingress, there is now a tendency to remove it and tile over. This is acceptable where adjoining roofs are level with one another, but should be avoided at **verges**. A skew will provide additional protection to the roof against the effects of driving rain.

Valley flashings are usually formed in lead, with the lead formed over a timber fillet to form a watergate *(see fig.11.15)*. Slates and tiles need to be trimmed at an angle and fixings can work loose. In tiled roofs the gap between lead and tile is often filled with mortar to prevent wind driven rain penetrating the gap.

Stairhead Glazing

When tenements turn a corner it becomes more difficult to get light into the close, hence the reason for **stairhead glazing** *(sometimes loosely termed 'cupola')*. However some common stairs were designed to be only lit by a toplight.

Invariably the stairs have an **open well** to allow the light to reach the lower levels. The glazing usually consisted of lapping panels of glass bedded in putty into timber glazing bars. Sometimes lead would be used to flash over these joints.

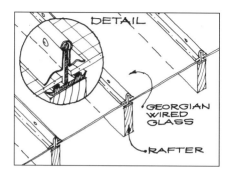

11.17 PATENT GLAZING

Renewing the Glazing

The modern equivalent of the puttyed **glazing bar** is referred to as a **patent glazing bar.** These are usually made from aluminium and the joint with the glass is made out of a neoprene or rubber gasket. The glass is usually installed in long lengths. The glass for the stairhead light should be georgian wired. This ensures that in the event of a fire the glass does not fall into the main escape route *(if the glass fell in it could create a funnel of air which would increase the spread of fire)*.

An alternative to **patent glazing** is to fit large rooflights. Those made by the Velux Company come with their own flashings and are double glazed *(this too should be replaced with georgian wired glass when fitted over close stairwells)*. The existing joists will need to be cut back and trimmed to allow for the wider opening - always make sure that the remaining joists are well stiffened as the new openings can weaken the roof structure.

In both options the builder will need to scaffold and plank out the close underneath before taking out the old glass. Access should be restricted and everyone warned when this is being done.

Broken glazing may keep water out for a period, but remember that broken glass could fall into the stairwell with serious consequences.

TV Aerials

If you are carrying out a full re-roofing project then it is advisable to have all the old aerials and leads renewed. During the works, the old aerials can be temporarily

fixed to the scaffolding. New common masts are fitted to appropriate chimneys and the new aerials fitted to them. Cables should be neatly clipped to the walls.

If the houses are being rehabilitated, it may be possible to fit a single **communal aerial** (or satellite dish) and service several closes. The leads can be taken down inside the flats in a conduit. A licence is required and the receiving unit requires a power supply.

Remember, the more people you can keep off the roof the better, and that includes TV repair men.

Chimneys

How a Chimney Works

Each fireplace has its own flue which carries smoke and exhaust gases upwards to be expelled at the chimney pot. Each flue is separated from any neighbouring flue so that smoke is not sucked back down into another fireplace.

If coal fires are in use the chimney has to be high enough to counteract any downdraughts from the roof. As a result of this, chimneys built on the front and rear walls need to be nearly as high as the ridge. Because of their height, the chimney may be strengthened with a metal stay. The stay is also designed as a ladder to provide access to the chimney sweep for cleaning the flues.

The large chimney between tenements will contain flues from both sides, hence the need to share costs of repair on these mutual chimneys.

When gas fires are connected to existing fireplaces, care needs to be taken to ensure the installation complies with the Gas Board requirements:

Flues will need to be tested and may need lined.

Flue terminals should be fitted rather than chimney pots.

All gas flues must be connected to the flue via a flue box, the old 'letterbox' openings are no longer acceptable. (The purpose of the flue box is to ensure that any soot or rubble which falls down the flue does not build up into an obstruction).

Drawing 11.18 shows how a chimney is constructed. Note that it is only the visible areas of the chimney stack that are built in stone, brick is used below roof level.

What to Look Out For

Because a chimney is exposed to the weather it can be expected to deteriorate more rapidly than other parts of the structure.

The stonework may also be attacked by sulphates from the soot or condensation of the gases in the flue. It is not uncommon to find the exposed stonework in a very soft and friable condition. In addition, check to see if the chimney is leaning or cracked or bulging. Any of these conditions will necessitate rebuilding the chimneyhead. If you are concerned about the stability of a chimney - no matter where it is - then you should report it to the Building Control Department. They have statutory powers which enable them to dismantle dangerous structures at very short notice.

The following items should also be inspected:

Stonework: if the stonework requires a lot of repair, then rebuilding is the best solution.

Open pointing: this can lead to water ingress and frost action on the stonework.

Copings: original copings are made from stone. They tie the chimney together and provide protection to the stonework. They may require replacement if worn or cracked. The chimney pots are held into the coping with cement flaunching. This flaunching deteriorates easily and may not be holding the pots safely.

Chimney pots: from the ground they may appear quite intact, but careful examination can show a multitude of hairline cracks.

The flashings: the junction between roof and chimney is always vulnerable to rain penetration. Old lead flashings may have deteriorated. Even with good flashings, moisture can find a way through the stonework. The timberwork around chimneyheads is particularly vulnerable to such moisture, often leading to outbreaks of rot.

The roofspace: always check the condition of brickwork under the chimneyhead as it can also be friable or open jointed, even to the point of allowing smoke to penetrate into the loftspace.

The flues: these can only be checked by carrying out a smoke test in each fireplace. As some fireplaces may be blocked this is not always possible. Blocked flues can sometimes be cleared by opening up the brickwork of the chimney and removing the obstruction. However if the flue bridges (see drawing) have collapsed, the work may be quite extensive. Provided only gas fires are connected to the flues, they may be repaired by inserting a flue liner (see dismantling and rebuilding procedure).

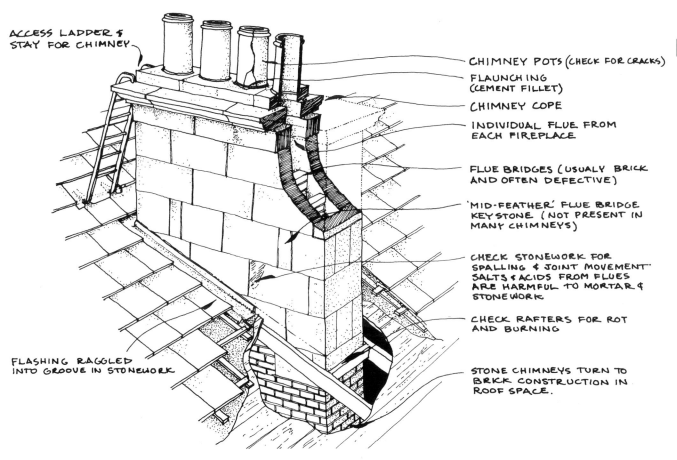

ACCESS LADDER & STAY FOR CHIMNEY

CHIMNEY POTS (CHECK FOR CRACKS)

FLAUNCHING (CEMENT FILLET)

CHIMNEY COPE

INDIVIDUAL FLUE FROM EACH FIREPLACE

FLUE BRIDGES (USUALY BRICK AND OFTEN DEFECTIVE)

'MID-FEATHER' FLUE BRIDGE KEYSTONE (NOT PRESENT IN MANY CHIMNEYS)

CHECK STONEWORK FOR SPALLING & JOINT MOVEMENT SALTS & ACIDS FROM FLUES ARE HARMFUL TO MORTAR & STONEWORK

CHECK RAFTERS FOR ROT AND BURNING

STONE CHIMNEYS TURN TO BRICK CONSTRUCTION IN ROOF SPACE.

FLASHING RAGGLED INTO GROOVE IN STONEWORK

11.18 CHIMNEY STACK

Rebuilding a Chimneyhead

Depending on the condition of the existing chimney, the choice between repair or rebuilding must be made. If the flues are good and the stonework sound then repairing will be the simplest option. However if there is any doubt about the chimneyhead, the opportunity of rebuilding it should be taken at the same time as other roof works.

In some cases chimneys are taken down and not rebuilt but capped off in the loft space. If the flues are relined, the new **flue liners** can be designed to terminate onto the roof through special **slate pieces** instead of going through a chimneyhead.

Chimneys are normally rebuilt in brick, but in some areas the Planning Department may insist on stone or brickwork rendered to match in with the existing stone colours. Where possible, cement render is best avoided on chimneys because of the high exposure. Facing bricks can usually be matched to the colour of stone quite well and it also avoids the problem of cracking and spalling render.

A good **coping** is essential. Old stone copings can sometimes be re-used, otherwise a new precast or in-situ concrete coping is required. Such a coping should be at least 100mm deep, have a weathered top surface and be **throated** on the underside with a sufficient overhang. (The throating is a small groove which prevents water drenching the face of the chimneyhead).

New pots are then fitted into the coping and pointed up.

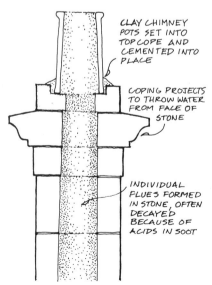

11.19 STONE CHIMNEYHEAD

Terracotta pots of all sizes are available. Gas cowls can be terracotta or aluminium.

Flashings are carried out in the same manner as flashings for skews.

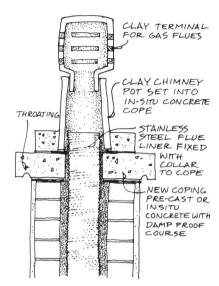

11.20 NEW BRICK CHIMNEYHEAD

Dismantling and Rebuilding Procedure

Because of the problems associated with dismantling and rebuilding chimneyheads, we now set down the stages of builderwork we see as good practice. These procedures may vary depending on circumstances, but in all cases the chimney should be properly scaffolded prior to demolition, and care taken to protect the public below through the use of boarded protection, enclosed chutes etc.

Stage 1. Inspect fireplaces. The builder contacts all residents and arranges to inspect the fireplace or gas fires where affected. Any defects in the gas fire or flue connection are noted. Defective fires cannot be reconnected unless they are repaired, so owners must be notified of this.

11.21 FIREPLACE INSPECTED

Stage 2. Remove gas fires. The builder disconnects all gas fires, labels them and puts them in a safe place. If a fire is defective, the owner should have it repaired.

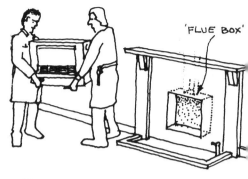

11.22 GAS FIRE REMOVED

Stage 3. Test flues. The builder lights a **smoke bomb** at the base of the flue and checks to see if the smoke is drawn up a single flue. If smoke appears out of two pots, the flues may be bridged and may need to be re-lined.

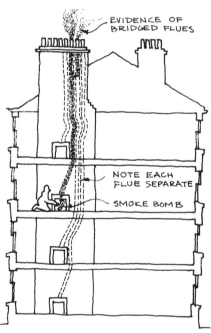

11.23 THE FLUE TEST

Stage 4. Seal fireplaces. The flue itself is sealed temporarily with old rags or newspapers. The opening is also sealed off as an additional protection against dust and soot. The builder should supply an alternative form of heating, but residents are expected to pay for their own heating costs.

64

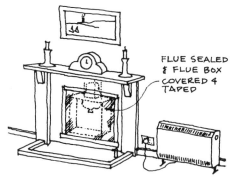

11.24 FIREPLACE SEALED

Stage 5. Seal flues. The chimney is scaffolded first and any necessary protection made to prevent rubble accidentally falling into the street. The chimney pots are removed, but before any dismantling begins, each flue should be sealed at the head of the chimney (*either using a sack filled with rags or Hardouns inflatable balloons*). This prevents debris from falling down into the flues and causing a blockage.

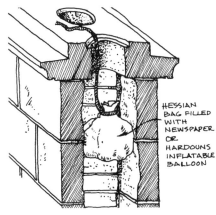

11.25 FLUE SEALED

Stage 6. Chimney lined. The chimney is now dismantled, usually to a solid brick base below roof level. If the flues are defective it is likely that re-lining will be necessary. Stainless steel **flue liners** (*trade name Copex*) are fed down each flue until they come out at the fireplace. A **flue box** should be formed to receive the flue (these may be preformed in metal or built in brick). Connections at top and bottom are made and the chimney is rebuilt. If flue liners are not used, then the flues should be retested once the chimney is rebuilt but before the coping is cast. The coping, pots and flashings are then installed.

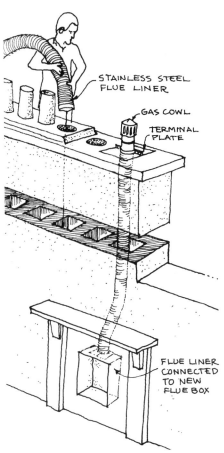

11.26 CHIMNEY LINED

Stage 7. Final test of flues. A final smoke test is carried out in the flue box. If it draws correctly then the gas fire can be re-fitted. This must be carried out by a CORGI registered gas fitter. Any damage caused to the room or fireplace should be rectified by the builder.

11.27 FINAL FLUE TEST

The Loft

The loftspace can reveal a lot about the roof structure and coverings, the plumbing system, insulation, skylights, electrics and ceiling condition. Its inspection should not be neglected.

Roof Construction

Scottish tenement roofs are normally constructed from:

Ceiling joists which span the tenement from front to rear and rest on a **wallplate** fitted to the top of the **wallhead.**

Rafters which are connected to the ends of the ceiling joists with a **pole plate** and connected to a **ridge piece** at the apex of the roof.

Hangers (*sometimes referred to as oxter pieces*), **struts** and **binders** give additional strength. **Cross ties** connect the two rafters (all these timbers are usually nailed and often notched together). Rafters and joists are trimmed around chimneys and skylights. The whole roof structure is considerably stiffened by the sarking timbers which form a fixing base for the slates.

65

What to Look Out For

Always check the condition of timbers next to the wallhead, and around chimneyheads and skews. These are the most vulnerable locations for dampness and hence rot outbreaks. Rotten pole plates or rafter ends can weaken the structure of the roof and cause outward movement which can even push out the stonework at the wallhead.

Check to ensure struts and ties have not been removed, as this can cause roof sagging.

Check any previous repair work, splicing, props etc, as these can suggest inherent faults in the roof structure.

Old slates and chimney debris are often left on top of joists in the loftspace. This can overload the joists and cause sagging to ceilings. Try to ensure the builder removes any rubbish if they are working on the roof.

If the ceilings in the top floor flats have to come down, then it may be worthwhile using a vacuum cleaner to remove the layers of soot and dust before dropping the ceiling.

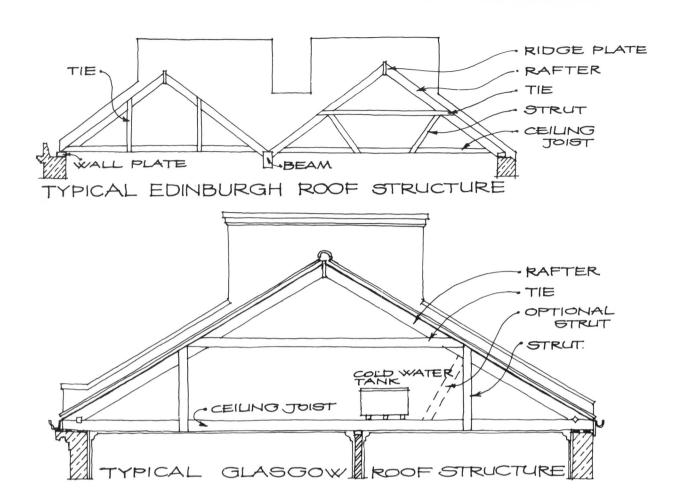

TYPICAL EDINBURGH ROOF STRUCTURE

Labels: TIE, WALL PLATE, BEAM, RIDGE PLATE, RAFTER, TIE, STRUT, CEILING JOIST

TYPICAL GLASGOW ROOF STRUCTURE

Labels: CEILING JOIST, COLD WATER TANK, RAFTER, TIE, OPTIONAL STRUT, STRUT

11.28 ROOF CONSTRUCTION

Firebreaks

Each tenement should be separated by a brick party wall extending up through the roof to the skew. If there are holes through this wall, fire could penetrate to the neighbouring loft. Always ensure any holes are bricked up. If there is a gable wall, check to see if there are any bulges or decayed brickwork as these walls may be poorly tied in to the rest of the structure.

Access

It is worthwhile having a good **catwalk** from the loft hatch to each water tank and to the skylight. Ply is better than chipboard as it is generally stiffer. Access to the roof should be from a small ladder at the base of the skylight. The skylight itself is generally made from cast iron, although velux-type rooflights are also used because of their ease of access. For additional security these skylights should have a padlock. The loft hatch is often poorly secured with a padlock. Remember that access to the top floor flats can be gained by punching a hole through the loft ceiling, so a solid loft hatch with a mortice lock is a worthwhile investment.

Electrics

The electric cables of the top floor flat will be visible from the loft. If the cables are old, the slightest movement can short the cables if the insulation is perished. If your roof is being replaced, it is likely that the exposed cables will be damaged and require replacement. It is not always fair to blame the builder if the old cable was needing replacement anyway.

Loft Insulation

The loft should be insulated with 150mm of glass fibre quilt (or a minimum of 100mm). Although this is strictly an individual improvement, it can be carried out under a common repair contract. Pipework and cold storage tanks should also be insulated, but do not insulate under the cold tank as the small amount of heat gain from the flat below can prevent freezing.

Some builders use a blown paper fibre insulant, however it can be difficult to control the depth of insulation consistently, and draughts at the perimeter eaves can blow the insulation away (being recycled, it is also a very green material). Because the insulation reduces heat loss, it makes the loft much colder and therefore more vulnerable to

condensation. This can be prevented by ensuring the loft space is well ventilated, preferably between the eaves and the ridge. Special slate vents can be fitted to replace a slate or a tile but the sarking will have to be cut.

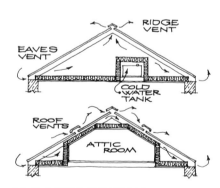

Labels: RIDGE VENT, EAVES VENT, COLD WATER TANK, ROOF VENTS, ATTIC ROOM

11.29 LOFT INSULATION

In some cases it may be necessary to install a vapour check to reduce the transmission of water vapour from the flat below. This is best done by applying the vapour check (foil backed plasterboard or polythene) to the ceiling below. Never lay polythene over the joists with insulation on top of it as this can allow condensation to wet the joists and cause them to rot.

66

The Structure of a Tenement

The structural integrity of a tenement depends on a number of inter-related elements tying together. Decay or movement of one or more of these elements can lead to major structural problems, so it's worthwhile knowing how these elements inter-relate.

Because of the height of the structure, the walls have to be thick enough to provide stiffness on the outer faces. Thick walls are also heavy, so adequate foundations are important. The outer walls are further stiffened by the internal walls and floors, and the close stair walls provide a stiffening core. The main floor joists usually span from front to back walls, and act as ties to stiffen the outer walls. These joists also need to spread some of their load onto the internal walls (usually the hall walls and bed recess area). The roof timbers provide further restraint at the top of the walls.

Damage to any of these areas can affect the stability and strength of the whole structure.

The External Walls

Most Scottish tenements were faced in stone, usually granite or sandstone. Although the wall would be about 600mm thick, the outer face might only be 150mm thick, the rest of the wall built from roughly cut stone or brick. In order to hold these two leaves together, **bonding stones** were built into the wall at regular intervals. Any void left between the two leaves of stone would be filled with smaller pieces of stone and mortar. These outer walls would be tied into the inner walls to give them greater stiffness.

Door and window openings would be spanned on the outside with stone **lintols,** but the inner section of the wall was often supported on timber lintols (called **safe lintols**). When there are large openings like shop windows, the stonework is supported on a **bressumer.** These beams are often timber, but are sometimes made of steel or cast iron.

Oriel windows are similar to bay windows, except they are cantilevered out from the wall at first floor level. This can be achieved by using steel or timber cantilevered joists or by corbelling the stone outwards and tying the wall back to a main **trimming beam.**

External walls are finished on the inside face with plaster on **lathes,** fixed to timber straps secured to the stonework.

The Internal Walls

The internal walls of the tenement are usually built in brick and may only be 120mm thick. These slender walls obtain a lot of their stability from the plaster coating on either side. They may be stiffened with door frames (which can span from floor to ceiling); by the action of other walls tying into them; and by the floor structure. These walls provide a box-like structure which gives rigidity to the tenement.

In some older tenements, the internal walls are made from timber **studs.** These walls may also be providing stiffening to the structure as a whole, and some could be **loadbearing.**

The Foundations

All **loadbearing walls** should be built on a foundation. The type of foundation will vary depending on ground conditions and local materials:

Large stones called padstones may be used, they are often not much wider than the wall thickness.

Brick footings will usually spread out so that the load of the wall is distributed.

Concrete foundations are found in more recent structures.

Timber foundations have been used in the past, usually to span boggy or unstable ground.

All bearing strata under the foundations will tend to compress under load. When a building straddles a mixture of different stratas, the building can sink unevenly, thus causing cracks.

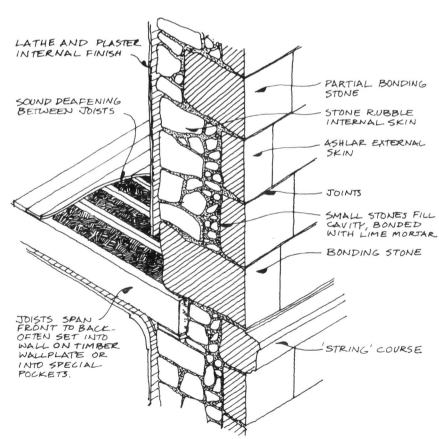

LATHE AND PLASTER INTERNAL FINISH

SOUND DEAFENING BETWEEN JOISTS

PARTIAL BONDING STONE

STONE RUBBLE INTERNAL SKIN

ASHLAR EXTERNAL SKIN

JOINTS

SMALL STONES FILL CAVITY, BONDED WITH LIME MORTAR

BONDING STONE

JOISTS SPAN FRONT TO BACK - OFTEN SET INTO WALL ON TIMBER WALLPLATE OR INTO SPECIAL POCKETS.

'STRING' COURSE

12.1 WALL CONSTRUCTION

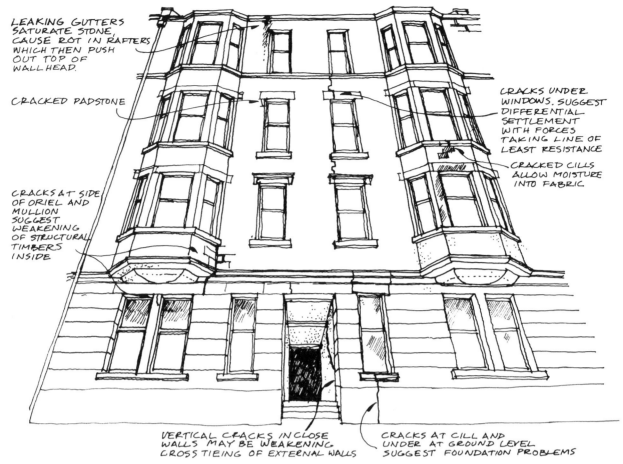

LEAKING GUTTERS SATURATE STONE, CAUSE ROT IN RAFTERS WHICH THEN PUSH OUT TOP OF WALL HEAD.

CRACKED PADSTONE

CRACKS AT SIDE OF ORIEL AND MULLION SUGGEST WEAKENING OF STRUCTURAL TIMBERS INSIDE

CRACKS UNDER WINDOWS. SUGGEST DIFFERENTIAL SETTLEMENT WITH FORCES TAKING LINE OF LEAST RESISTANCE

CRACKED CILLS ALLOW MOISTURE INTO FABRIC

VERTICAL CRACKS IN CLOSE WALLS MAY BE WEAKENING CROSS TIEING OF EXTERNAL WALLS

CRACKS AT CILL AND UNDER AT GROUND LEVEL SUGGEST FOUNDATION PROBLEMS

12.2 WHERE TO LOOK FOR CRACKS

Structural Problems

Cracks and bulges in a wall can be caused for a variety of reasons, including: *thermal movement; differential settlement; structural decay; and vibration.* A crack in the wall does not necessarily mean that the wall is unstable, so care has to be taken to interpret the cause of any cracking or distress.

Most cracks will appear early in the building's life. But some will remain active, suggesting continued movement or settlement. The rate at which cracks move can be measured with a **tell-tale.** This simple device is placed across the crack and secured to the walls on either side. Readings are taken at regular intervals.

If you are concerned about the structural state of your tenement you would be well advised to appoint a civil engineer or an architect to carry out a structural survey.

If you are surveying it yourself, it is a good idea to make a diagrammatic sketch of the elevations first, then mark on the location and extent of the defects. This chapter explains some of the more common problems.

Localised Cracks

Many cracks are localised to individual stones or small areas and do not affect the overall structure of the building.

Cracked lintols suggest overstressing, and may be caused because the safe lintols behind are rotten. Because of this internal failure, the load is transferred to the outer face, thus overstressing the lintol.

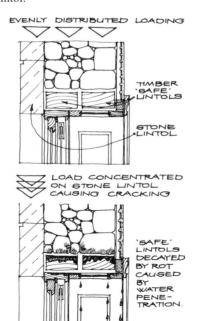

EVENLY DISTRIBUTED LOADING

TIMBER 'SAFE' LINTOLS

STONE LINTOL

LOAD CONCENTRATED ON STONE LINTOL CAUSING CRACKING

'SAFE' LINTOLS DECAYED BY ROT CAUSED BY WATER PENE- TRATION.

12.3 SAFE LINTOLS

Cracked bearing stones (the stone supporting the lintols also called padstones), may be caused by overstressing or poor stone bedding.

Bulges

In some cases the outer wall may bulge outwards or inwards. This is usually caused by lack of tying between the floor joists and the wall structure. Timber joists or window lintols may be rotten at this area so an internal inspection is always recommended.

If the stonework is leaning outwards at eaves level, it is likely that the roof timbers have moved and are pushing the stonework out. Rot at the eaves is the most likely suspect.

Bulges are also found on ground floor close walls, usually where the wall is not loadbearing as such walls are very tall and slender and have no added weight to keep them restrained.

Bulges and leaning walls can be checked by dropping a plumb line from the top floor windows.

Separation Cracks

These often appear on stair landing walls at the junction of the close wall and the outside wall. This area is vulnerable to cracking because the tying of the internal

half brick walls is poorly done (usually with a single brick tie every fifth course).

Because the outer wall is bearing a much greater load we may expect it to sink at a different rate from the internal wall.

Generally these cracks are not serious, but where the crack is wide, or if there is **differential movement**, then structural repair may be necessary.

Floors Off-Level

Sometimes the outside stonework appears perfectly level but the floors inside are off-level. This can be simply checked with a marble or a spirit level. The squareness of door frames should also be checked. If floors are off-level it may be caused by rot in the joists or the sinking of internal support walls. In ground floor flats it is not uncommon to find that under the floor, holes have been knocked in the walls to provide a crawl access for an electrician. Such damage can lead to movement of the walls above.

It is also possible that 'improvements' have been carried out in the flat below and structural walls (like the bed recess wall) have been removed.

Differential Settlement

In the case of differential settlement, the cracks may be through the structure of the wall and extend some distance up it. Cracking will take the line of least resistance so watch out for the area above and below the windows. Also look out for cills and lintols which are off level or cracke; the tenement wall has probably sunk at a different rate to the adjacent section. The main forms of ground movement which cause this are as follows:

Excessive moisture in the ground can cause clay subsoil to expand and push upwards.

Tree roots can also push walls up and crack foundations. In very dry weather the tree can take all the moisture out of the ground causing the subsoil to settle. One advantage to Scotland's weather is that the subsoil rarely gets this dry.

Burst pipes in the basement areas can gradually wash away the loose fill around the foundations causing localised settlement.

The nature of the ground itself. If the underlying strata varies then settlement will occur at different rates.

Old mineworkings are rarely backfilled. The timber props eventually rot away and the ground collapses into the void.

If foundations are suspected then trial pits may have to be dug to inspect the condition of the footings. If the strength of the ground is suspect, then a test bore through the ground may be needed. The core sample taken will tell what type of subsoil or rock exists.

Structural Repairs

There are a variety of solutions which must all be tailored to meet the circumstances of each particular problem. The following is a summary of how some of the defects above may be repaired:

Local Crack Repairs

Small cracks can usually be repaired by cutting into them and filling with a weak mortar mix, coloured to match the stonework.

Cracked lintols will require strengthening: in the past they were often repaired by the insertion of a flat steel bar underneath. This gave little support to the lintol. It is now more common to insert a galvanised angle iron underneath it, or to replace the lintol completely. New lintols can be made of stone or cast stone (a concrete substitute). Some cracks can be repaired by stapling - that is, fitting a stainless steel or rustproof staple across the crack and pointing up to disguise the staple.

Tying the Building

Some bulges and wall movements (separation cracks) can be restrained by fitting ties into the building. A steel beam or spreader plate is fitted on the face of the building where the problem is, and steel rods or straps are taken into the building to be connected to internal walls or floors. In some cases these ties may be taken through the floor from front to back, thus stiffening the whole structure. Although such work can solve the structural problem, the visibility of the steelwork can suggest to prospective purchasers that there is a structural problem. It may therefore be worthwhile to disguise these repairs by:

Sinking the steelwork into the stone and rendering over it.

Fitting a fake cornice over the steel to make it appear like a string course or

building feature.

Fixing the ties to the internal face of the building with resin anchors.

"AYE, THEY MADE A GRAND JOB OF TYING THAT TENEMENT"

Wall Rebuilding

If the walls are badly out of plumb or decayed, there may be no alternative but to rebuild a section of the wall. Such work usually necessitates decanting the residents because of the need to secure temporary support to the remaining structure. The new wall can be rebuilt in stone, cast stone, concrete block or usually brick. Facing brick can be chosen to match the colour of the stone, or the brick can be rendered with a stone coloured mortar.

Bed Recess Walls

These internal walls are structural as they carry some of the weight of the floors above. If they are removed, a new lintol will need to be fitted to carry the weight of the floor above. If the adjacent walls are very slender, new brick piers may need to be built from foundation level in order to take the increased load on the remaining walls.

Beam Replacement

Bressumer beams which span wide openings, at pends, shopfronts, or oriels may need to be replaced because of wood rot. Great care has to be taken during this type of work as the weight of the tenement will need to be securely supported while the operation is carried out. Temporary beams (needles) and props are fitted through the wall to provide this support and then hydraulic jacks are used to ensure the load of the wall is transferred to the temporary props.

If this is not done, further settlement can occur during the replacement work.

Foundation Repairs

In some cases it is the foundation itself which needs repaired, but in most cases of differential settlement it is the ground underneath the foundations which needs reinforced in some way. The following are some of the main methods of repair:

Underpinning

This can be used to replace a defective foundation or to provide a stronger foundation which can spread the load of the tenement over a wider area. The work is carried out in stages. Pockets are dug about 1200mm wide under the old foundation or wall footing and a new reinforced concrete foundation is laid. Once it is set, the adjacent pocket can be dug and another section cast into place, tying into the reinforcement of the first section.

Pile Stitching

Short concrete **piles** are driven in at an angle under the foundation from both sides of the wall. The load of the wall is then supported on the new pile stitches which spread the load over a much wider area.

Grouting

This is usually carried out on an area basis as it is designed to fill voids under the foundations (i.e. old coal workings). Holes are drilled under the foundation area and a **grouting material** pumped into the ground to consolidate it.

When Things Go Wrong

Once again we can only advise you to ensure you use a civil engineer to check the structure and to supervise any major structural works. A Building Warrant will be required for most of the structural repairs mentioned here. **If you are in any doubt, or notice some sudden movement or cracking in your building, then don't hesitate to call out the Building Control Department.** They have special powers which allow them to demolish unsafe structures or to make emergency repairs of a structural nature.

Stonework

In the 17th century in Scotland, timber was extensively used in housing construction, but civic pride and a growing awareness of fire risks led to the increased use of stone. Local councils even gave tax free allowances to encourage stone building.

In the 18th century walls would be made from uncut boulders, with **freestone** only used around corners (**quoins**) and at window openings (**rybats**). The boulders would be **harled** over. The Georgians and early Victorians tended to use stone from local quarries because of the difficulty of transporting it. This stone was usually cut, finished and coursed on the front elevations (**ashlar**); on the rear elevations a **squared rubble** stone would be used, sometimes coursed but not always.

In Glasgow, local quarries were Kenmure, Westfield and Giffnock. When this cream sandstone was exhausted in the 1890s it was replaced with the red sandstones from Dumfries and Lockerbie.

In Edinburgh, carboniferous sandstone came from Craigleith and Broughton up until 1900. This sandstone was generally of better quality than the softer sandstones of Glasgow.

In Aberdeen, granite was used from the Rubieslaw quarry, not closed until 1971.

What to Look Out For

If you are carrying out a major repair scheme to your tenement, the architect or surveyor will carry out a survey of the stonework from the ground. An elevational drawing is prepared to show the extent and nature of the stone decay as well as what repairs need to be carried out on the stone. Unfortunately, not all defective stones can be identified (or analysed) from a ground based inspection, even using binoculars. Once the scaffolding is erected, a more detailed stone survey will be carried out. Each stone may be inspected by lightly tapping with a masonry hammer to ensure its solidity.

The decay of stonework can be caused by a number of different reasons:

Weathering

Sandstone is by nature porous and will take in moisture and release it depending on the weather. The continuous action of wetting and drying can wash out the material which binds the individual grains of sand which form the sandstone (referred to as grain boundary clay). A stone which is very porous may be more susceptible to the action of frost from within, causing the stone to flake off. Stonework which is poorly protected, or more exposed (see chimneyheads on page) will receive an excess of moisture which speeds up this weathering process.

Soluble salts

Most stones and mortars contain salts, which when dampened can mix together and crystalise. These new salts then expand, like the action of frost, and disintegrate the stone. For example, chimneys are vulnerable because of the soot content in the flues. The salts in the soot can leach through the stone and lead to staining on the outer face.

Pollution

Stonework can be attacked by chemicals in the atmosphere. Carbon dioxide has the effect of dissolving calcium carbonate within the stone which in turn destroys the structure of the stone. Pollution also dirties the stone, indeed a newly cleaned tenement can expect to darken within the first few months due to soot deposits and diesel exhaust. Diesel fumes are now one of the more serious threats to stonework in city areas.

ASHLAR

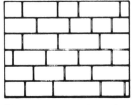

POLISHED ASHLAR COURSE

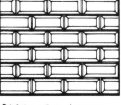

RUSTICATED ASHLAR – 'V' JOINT.

BANDED ASHLAR

ASHLAR FINISHES

  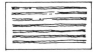

POLISHED ASHLAR   DROVED ASHLAR   STUGGED ASHLAR   BROACHED ASHLAR

SQUARED RUBBLE

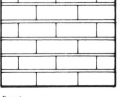

REGULAR COURSED SQUARED RUBBLE

COURSED SQUARED RUBBLE

RANDOM RUBBLE

UNCOURSED RANDOM RUBBLE

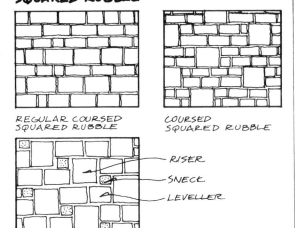

- RISER
- SNECK
- LEVELLER

UNCOURSED SQUARED RUBBLE WITH SNECKS

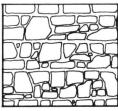

COURSED RANDOM RUBBLE

13.1 STONEWORK FINISHES

Workmanship and quality.

Not all stone is of the same quality. Even stone from the same quarry will have a variation in quality. Cheaper stones were often used to build cheaper properties. Sandstone, being laminated, needs to be bedded on the correct plane. Incorrectly bedded stones will be prone to decay because the natural movement of the laminations, originally held under great pressure are now unrestrained.

13.2: Extensive decay of stonework

Stone Repairs

There are two main methods of repairing defective stonework: by **indenting** with new stone; by **mortar repairs**.

Indenting

This involves cutting out the defective stone and replacing it with a new stone selected to match the existing. It is generally always required in Grade A

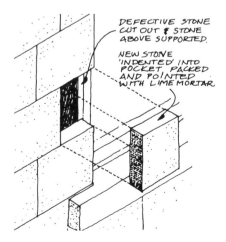

13.3 INDENT REPAIRS

listed buildings and may be considered a planning condition in some conservation areas.

The complete length of the stone need not be removed, as long as the new joint can give a bonding pattern. The new stone should be anchored into the old stonework by means of stainless steel fixing cramps or non-ferrous clamps. **Dowels** can also be used, bedded into the stonework and fixed with **epoxy resin**.

This type of work should only be carried out by an experienced mason. It is a lot more expensive than a mortar repair, but is a lot more permanent.

Mortar Repairs

Mortar repairs (also loosely referred to as **plastic repairs**), are carried out using a mixture of **cement, lime** and sand (usually about 1:2:9 proportions). The cement should be **sulphate resisting.** The colouring should be obtained by selecting the appropriate sand colour rather than depending on mortar dyes which fade with time. In a mortar repair, the decayed stone is cut back to a sound surface. It is then wetted before applying the mortar. In some cases, it may be necessary to fix copper wire reinforcement to the old stone surface first in order to provide a good key for the repair. The mortar mix is usually applied in two coats, and should never be **skimmed** over the stone. The finished surface can be given a texture to match the existing stonework.

There is also a special restoration mortar called Keim Restauro Grund which contains natural stone substitutes and is meant to be more compatible to the original stone. It can also be pre-coloured to match the existing stone *(see address for Keim in Appendix 6).*

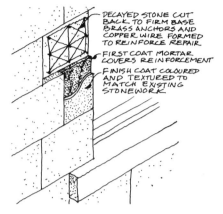

13.4 MORTAR REPAIRS

If the colouring has been unsuccessful, or if the stonework is badly stained, it is possible to apply a special resin coating to the surface of the stone. The treatment includes bonding a coloured sand to the surface of the stone (trade name linostone, see page 74).

Common Faults

The quality and workmanship of mortar repairs need to be carefully supervised. Some small builders use plasterers who have no experience in working with stone, to carry out mortar repairs.

The commonest faults are as follows:

Cement repairs with no lime and a high cement content giving too strong a mortar mix. The repair is denser than the surrounding stone or backing and will eventually loose adhesion or damage the surrounding stone.

Rapid drying out: either as a result of too strong a mix or premature drying in very hot weather. Hairline cracks develop in the mortar which allows moisture to penetrate and leads to frost action behind the repair.

Mortar repair too thick: if the mortar build-up is too thick and is without reinforcement, cracking can develop.

Mortar repair too thin: the repair should be at least 25mm thick. If the mortar is skimmed to a feather edge over the stone, it will eventually crack and loose adhesion.

Stone Features

Cornices etc
Cornice stones, **parapets** and window **pediments** are commonly found decorating Victorian tenements. Such cornices and projections protect the external stonework, but as a result, receive more weathering than the surrounding ashlar. These mouldings can be reformed using mortar repairs although reinforcement is often required. If the stone is projecting some distance it may be worthwhile covering the exposed top surface in lead for additional protection.

Mullions
Stone mullions *(the vertical pillar of stone found between some windows)* which are cut from the wrong bed have a tendency to delaminate. A mortar repair in some cases may be only cosmetic as the structure of the stone is so badly damaged. The mullion can be replaced, usually with a stone or cast stone replacement. The new mullion has to be restrained with non-ferrous straps or dowels fixed at the head and cill.

Cills
Stone cills can also be repaired using mortar. If badly decayed a new cill (stone or cast stone) can be inserted. Cracks in the cill will allow water to penetrate and could cause rot outbreaks. If the crack is fine, it can be repaired with **lime mortar grout**, otherwise it may be best to consider replacing the cill.

Stone Pointing

Mortar for bedding and pointing sandstone walls is usually made from a mixture of sand and lime. If the pointing is missing, water can cause more damage to the stone. The type of pointing and treatment will depend on the type of stonework. Remember that a lot of damage can be done by badly pointing a building. Indeed provided the joints are not open to the weather, the less pointing you do the better.

Ashlar Wall Pointing
The front elevation of evenly coursed ashlar may have been pointed with a **white lime putty.** These thin white courses are difficult to repoint because removal of the old lime putty can damage the surrounding stone. If the thin white pointing is to be reinstated, lime putty is used and the stonework is protected with masking tape. Wider joints can be raked out and repointed with a sand/lime mix, or a weak mortar mix (1:2:8). It may be preferable to tone the pointing material to match the stonework.

Rubble Wall Pointing
The joints between **rubble stones** are generally wider. Old pointing should be carefully raked out to a depth of at least 25mm and cleaned and wetted before repointing *(power tools should never be used to remove old pointing)*. The pointing mix will be similar to the ashlar pointing and the finished joint should be slightly recessed behind the **arris** of the stonework.

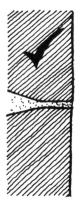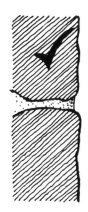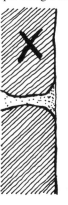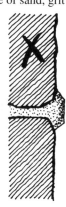

THIN JOINTED ASHLAR - MORTAR SLIGHTLY RECESSED OR POINTED WITH LIME PUTTY MORTAR

RUBBLE POINTED WITH SLIGHTLY RECESSED MORTAR, BRUSH FINISHED.

AVOID FLUSH POINTING UNLESS USING LIME MORTAR

NEVER POINT PROUD OF STONEWORK

13.5 STONE POINTING

Stone Cleaning

The cleaning of sandstone tenements has been actively encouraged in Glasgow, whilst in Edinburgh it has generally been discouraged. There are arguments to support both policies.

Recent research has shown that all cleaning is in some way harmful to the stone. All cleaning tends to remove the protective outer skin and open up the pores of the stone thus making it more vulnerable to weathering. As it is hard to know the long term effects of this process, the preferred option is to avoid cleaning. On the other hand the cleaning of tenements can increase confidence in an area as the cleaned stonework is a visible sign that tenements have been improved. Futhermore people begin to look at the tenement in a new light and like what they see.

If you are having your tenement cleaned as part of a repair contract, your architect will have to apply for planning permission. This may set down certain conditions on the type of stone cleaning allowed.

Stone Cleaning Methods
The main methods are as follows:

Dry Sand Blasting: this is an abrasive technique used extensively in Glasgow up until 1983. It is now restricted in Glasgow because of the health risk caused by the dust it creates. In the hands of an unskilled operator, the stone could be badly damaged, but if used correctly it can be one of the less damaging methods.

Wet Grit Blasting: another abrasive technique using a mixture of sand, grit and water. The water reduces the amount of dust created but stonework can be damaged in the same way as dry grit blasting. The mixture of grit and water creates a slurry which in itself may need to be cleaned from the stone. Different types of grit are now available which are meant to cause less damage to the stone, although recent studies suggest that the pressure it is hosed out at is the biggest factor in controlling damage.

Washing: water can be jetted over the stonework at high or low pressures and the method is often used to clean limestones. Remember that high pressure water jetting can be used to cut steel. If the water pressure is not closely regulated, the stone can be damaged. Excessive wetting of the stone can speed up the weathering process and open up the pores of the stone.

Chemicals: some chemicals can cause more harm than good to the stonework. **Hydrofluoric acid** is most commonly used on sandstones. It requires skilled operators to apply it and must be thoroughly washed off the building after a set period. Recent studies have cast doubt over this method of cleaning, because the acid is never completely removed by washing.

Neolith 60 is a paint and varnish remover which is non-caustic but must be thoroughly washed off. They also recommend that windows and glazing should be protected with the application of Tak Pro Peel Latex protection film. Keim also make a non-caustic stone cleaner type N which is claimed to not harm the stone.

Painted Stonework: It is never advisable to paint stonework because it can prevent the material from 'breathing'. Shops which have ground floor areas painted may be best left alone or repainted. Oil paints can be removed with a non-caustic paint remover or even burnt off carefully. It is rarely possible to return the stone to its natural state so in some circumstances a surface decorative treatment such as Keim paint may be applied.

Stains: once the stonework has been cleaned, stains may become visible. These were hidden by the layers of grime and dirt. Stone staining may be due to salts coming to the surface of the stone after it has been dried out. Although unsightly, the staining should tone down after a few years. It is always better to leave the natural stone wherever possible rather than trying to seal the surface for decorative reasons.

Linostone Treatment

Once cleaned, the variation in stone colour will be visible and will generally enhance the look of the building. Linostone is a proprietary decorative resin coating which is sometimes applied to stone repairs or stained stones to make them match the colour of the cleaned stone. It is generally applied in three coats:

1) a primer coat.

2) linostone backing coat (white in colour).

3) linostone base coat which whilst wet has a coloured sand cast onto it to provide the colour and texture.

White pointing joints can be simulated by placing a thin masking tape over the white base coat. The linostone is then cast and the tape carefully removed. It is possible to get a wide variety of linostone colours. Unfortunately the life expectancy of this kind of decorative treatment is limited, particularly when poorly applied, and is best avoided if possible.

13.6 Defective resin coating treatment

What to Look Out For

If your tenement is about to be cleaned, make sure all the windows are firmly sealed with masking tape inside and out. The glass and timber should also be covered with polythene *(sand-blasters have been* known to etch glass accidentally).

If water is used in the cleaning, try to ensure hoses are not left to saturate walls. This can lead to water penetration, frost attack and even rot problems.

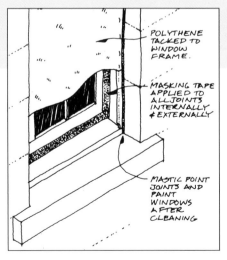

13.7 WINDOW PROTECTION

74

The Close

The close is a unique feature of the Scottish tenement. It is shared by all the people who live in it, and is walked through by all the visitors. It is neither public nor private. All the different elements *(the floor, walls, ceilings, stairs, services, doors and windows)* require cleaning and maintenance, and new owners or tenants should be made clear of their responsibilities at the outset.

The Close Floor and Landings

The close floor, from the entrance, is usually slabbed in stone. In a majority of cases, the **rising main** and drainage pipe is located under these slabs.

The landings outside flat entrances are usually supported on timber joists, and the space between joists packed with ashes. Half-landings are usually supported on steel bearers.

What to Look Out For

If the floor sounds hollow, there may be a basement or void underneath. The stone slabs may be spanning the close width or supported on steel beams. A closer inspection may show how weak the floor structure is.

If there is a drain blockage, it may occur under the close, necessitating renewal of the floor. Leaks in the water supply pipe *(rising main)* under the floor can go unnoticed for ages. Leaking water can eventually wash away the surrounding soil (and foundations!)

Frequent cleaning of the stair can allow water to seep between stones on the landings, and gradually rot any timber supports.

A New Close Floor

There are a number of options, starting with the simplest:

1) Lift the existing stone flags and relay them. You may have more joints, but you are left with the benefit of a hard wearing surface.

2) Lift the floor and relay with concrete paving slabs. If a good slab is chosen this can be effective. The slabs can be treated with a sealant to make them easier to clean.

3) Screed over the existing floor with a polymer concrete. This is usually about 40mm thick and will give an even surface. It can be difficult to clean and is vulnerable to staining. It can also be treated with a floor paint to make it easier to clean.

4) Screed over existing floor with compound resin material. Often referred to as a Veitchi floor, a number of suppliers will lay their version of it. The resin material can be taken up at the sides to form a skirting, making the floor easier to clean. Areas around close entrances should be tiled for added weather protection.

5) Lifting the old floor and relaying a new concrete floor. It should incorporate a DPC, particularly if it is to be tiled. Expansion joints may be required at the perimeter of the close walls, and at other areas (without them cracking of the slab may occur).

6) As above, but with the addition of clay quarry tiles throughout the close. Special tiled skirtings are also available for easier cleaning.

The Close Stair

In the 18th century, staircase access to upper flats would have been external and at the rear, with access through a pend. When tenements were built higher than 2 stories, the stair, usually circular, was enclosed in a tower. Over the centuries, the stair has come further into the tenement, eventually to be part of the tenement with windows to the rear. At corners, and in some closes, an internal toplit stair would be used. Architects of the 1890s even made the stair a feature by bringing it to the front of the building, a vogue which is still in use.

In outline, the stair types most commonly found are as follows:

Stairs with a central spine wall: if the staircase is modern, it is likely to be in-situ concrete and will span in one piece between the landings; if the stairs are made from individual units for each tread, then they will span between the spine wall and the close wall.

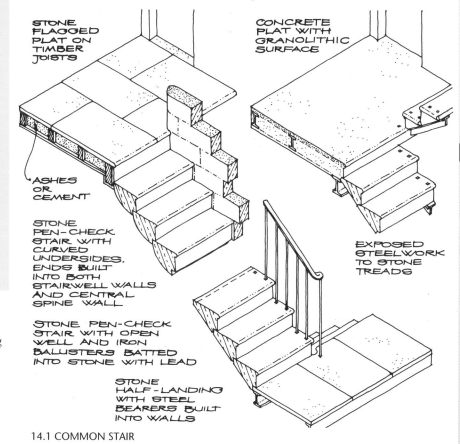

STONE FLAGGED PLAT ON TIMBER JOISTS

CONCRETE PLAT WITH GRANOLITHIC SURFACE

ASHES OR CEMENT

STONE PEN-CHECK STAIR WITH CURVED UNDERSIDES. ENDS BUILT INTO BOTH STAIRWELL WALLS AND CENTRAL SPINE WALL

STONE PEN-CHECK STAIR WITH OPEN WELL AND IRON BALUSTERS BATTED INTO STONE WITH LEAD

STONE HALF-LANDING WITH STEEL BEARERS BUILT INTO WALLS

EXPOSED STEELWORK TO STONE TREADS

14.1 COMMON STAIR

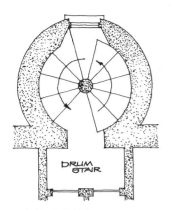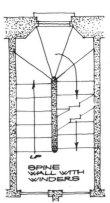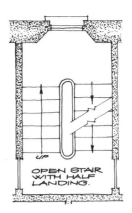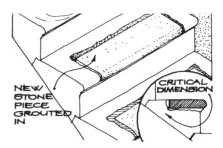

14.4 INDENTED REPAIRS

14.2 STAIR TYPES

Cantilevered pencheck stair with open balustrading. Each individual stairtread is shaped to interlock into the one above. The treads are built into the close wall. In some cases there may be a beam supporting the unsupported edge.

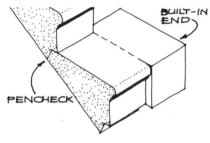

14.3 PENCHECK STAIR

Curved and 'open' pencheck stairs which provide an open well allowing toplighting of the close.

Landings would often be constructed by placing large stone slabs over timber joists (steel was only rarely used here).

Retreading Stairs

Naturally enough, the first flight of stairs receives more wear than the top flight. It may not be essential to renew all the treads, but it is often the best solution as individual stair retreads can lead to uneven step heights. There are a number of different treatments:

Polymer concrete retread – often the cheapest solution, it involves screeding over the existing **treads** and **risers** with a **polymer concrete** (this allows the screed to be relatively thin, about an inch). The edges (**nosings**) of the steps have to be made square. The finished surface is meant to be non-slip, but the texture of the surface can make the close difficult to clean. A concrete sealer or epoxy paint can help slightly.

Indenting – if only a number of treads are worn it may be better to indent the repair. A **pocket** is cut into the tread over the worn area and a new piece of stone is grouted into the pocket. Alternatively the

pocket can be filled with a polymer resin/stone chipping mix.

Composition flooring - this is a resin based material applied over the steps, floors and landings. (It is commonly called Veitchi-floor after the company of that name). The material is very flexible and comes in a variety of coloured 'marbled' finishes. Cleaning the floor is very easy. Stair nosings can be made with brass or plastic nosings, and non-slip inserts are available for the treads.

Balusters

Most Victorian balusters are made from cast iron, often in a decorative shape. They are set into a hole in the stone tread and caulked with molten lead. Each baluster is joined to a small steel rail at the top. A timber handrail is then fixed over this rail and screwed into position from the underside of the metal rail.

Broken balusters can be replaced with new cast iron balusters, but the cost is high. Welded repairs are possible in some cases. A shaky flight of balusters can sometimes be secured by fixing metal ties between one flight and the next.

What to Look For

The staircase and surrounding walls provide a **structural core** to the tenement. If one flight collapses it can bring down other flights, and even push out the back wall. Curved and open pencheck stairs are perhaps the most vulnerable. Check the following:

Subsidence - if stairtreads are off-level it can point to differential settlement of a close wall. This may be caused by poor foundations or poor ground conditions. A leaking water main or drain underneath the close can be enough to gradually wash away parts of the foundations.

Stairtread cracks - treads can be cracked by heavy or sudden loads. If the crack is an isolated one it can usually be repaired by fixing stainless steel staples across the crack. If it is serious or extends to more treads,

then steel strengthening or rebuilding may be required.

Gaps between penchecks - movement of one pencheck tread can leave the rest of the flight virtually unsupported: check the width of the joints between individual treads, they should be tight.

Loose balusters - on a curved and 'open' pencheck stair the balusters and handrail give added strength to the construction. Loose or broken balusters should therefore be repaired promptly.

Separation cracks - these occur at the junction of close walls and the rear wall. It happens because the close walls are rarely keyed into the rear wall very well. Small cracks are rarely serious, but larger cracks may necessitate the fitting of structural ties

between the close wall and the rear wall.

Rot outbreaks - where landings are built over timber joists, excessive washing of the floor can lead to rot outbreaks under the landing slab. It is a fairly rare occurrence.

CAUTION
Curved and open ended pencheck stairs are vulnerable to collapse through cracking and movement of the individual treads. Because of this, Building Control will usually advise that the stairs are reinforced with additional steelwork on the underside of the flights.

In all of these cases we would advise you to contact a structural engineer, architect or building control officer if you are in any doubt.

Handrails

Timber handrails can usually be repaired, small holes and indentations being filled with a stopping compound. If your close has a central spine wall with no handrail, then a new one can be fitted to the wall with metal brackets. If the brickwork is poor, it may be advisable to fix a timber base plate to the wall first, then fix the metal brackets to the timber. New metal handrails can be finished by applying a plastic cover over the metal.

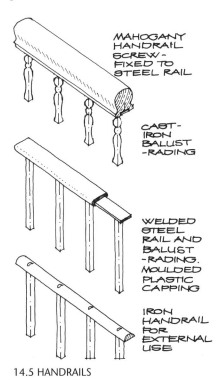

MAHOGANY HANDRAIL SCREW-FIXED TO STEEL RAIL

CAST-IRON BALUST-RADING

WELDED STEEL RAIL AND BALUST-RADING. MOULDED PLASTIC CAPPING

IRON HANDRAIL FOR EXTERNAL USE

14.5 HANDRAILS

Close Doors

Many closes in Edinburgh still retain their original eighteenth century close doors. They were solid panel doors with an offset pivot hinge which ensured the door stayed closed. The door latch could be opened by a lever on each landing which pulled a concealed wire connected to the latch mechanism. In Glasgow and Clydeside close doors were a rarity until improvement work began to be carried out in the 1980s.

The advantages of fitting a close door are twofold:

> *They provide additional privacy to the close. A door entry system will increase security.*

> *They prevent excessive draughts and heat loss in the close. A lot of heat can be lost through thin close walls into the close. Fitting a close door front and back raises the temperature of the close slightly thus reducing this heat loss.*

LEVER TO EACH LANDING

LEVER OPERATES LATCH TO FRONT DOOR

14.6 CLOSE DOOR IN EDINBURGH

As the fitting of a new close door is not a 'repair', it is not strictly eligible for a repair grant, unless you are carrying out a full repair scheme. The new door must have everyone's approval first - you cannot force your neighbour to share the cost of the new door. Rear close doors are sometimes installed as part of an environmental improvement scheme.

A New Front Close Door

There are many different designs which you can choose from. If you live in a conservation area you should first of all approach the Planning Department for advice on the most suitable design. The door can be:

Framed and panelled like an old fashioned door.

Framed and fully or partly glazed.

Solid core doors; some types are vulnerable to movement from moisture, and as such cannot be recommended for close doors.

The door will normally be a **single leaf door**. There is a vogue for fitting double leaf doors but the close width is rarely wide enough to provide a suitable size for the two doors.

A New Rear Close Door

These are generally made lockable so that access through the close, for people taking shortcuts, is prevented. The door can be solid, glazed or panelled. It is common to have a framed door with vertical tongued and grooved panels in the bottom section and a glazed top section to allow light into the close. If there is a lot of vandalism in the area, a metal grill instead of glass will be quite effective.

Ironmongery

The door can be locked with a number of devices:

Mortice lock: if there are a lot of new close doors around the backcourt then it makes sense to have them **master keyed**.

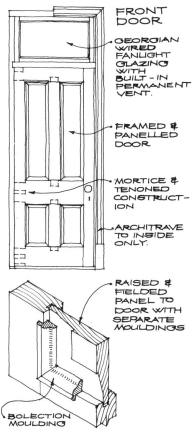

FRONT DOOR

GEORGIAN WIRED FANLIGHT GLAZING WITH BUILT-IN PERMANENT VENT.

FRAMED & PANELLED DOOR

MORTICE & TENONED CONSTRUCTION

ARCHITRAVE TO INSIDE ONLY.

RAISED & FIELDED PANEL TO DOOR WITH SEPARATE MOULDINGS

BOLECTION MOULDING

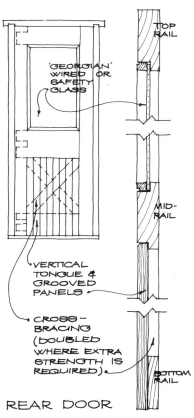

TOP RAIL

'GEORGIAN' WIRED OR SAFETY GLASS

VERTICAL TONGUE & GROOVED PANELS

MID-RAIL

CROSS-BRACING (DOUBLED WHERE EXTRA STRENGTH IS REQUIRED).

BOTTOM RAIL

REAR DOOR

14.7 CLOSE DOORS

Remember that if your bins are collected through the close the cleansing department will need access. They usually request 4 sets of keys.

Throw bolt: simple and effective, but it will not prevent people entering at the front and making their way through the back door.

Star key: the star key is common and easily replaced, and it will prevent access like the **throw bolt**.

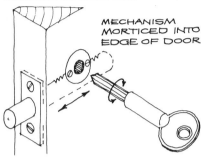

MECHANISM MORTICED INTO EDGE OF DOOR

14.8 STAR KEY

Other items of ironmongery include:

In areas of high vandalism, knobs can be stronger than lever handles, but the latter provide easier opening.

A door closer should ensure the door is kept closed and should prevent banging.

A hook and eye are useful in keeping the door open when required.

Kick plates and finger plates will protect vulnerable areas of the door from wear. They can be secured with non-returnable screws.

Hinges should be rustproof as should all screw fixings, and three hinges reduce the likelihood of warping of timber.

Door Closers

There are a variety of types:

Overhead Door Closer

These are fitted at the top of the door. They should ideally be bolted through the door otherwise they can be stolen easily. The closer provides a **closing** and **checking action**. The speed of closing can be controlled. The checking action is meant to slow the door down before it hits the frame and prevent banging. Some closers have a facility which allows the door to remain open when opened to 90°.

Floor Spring Closers

These are generally the most robust and less liable to damage from theft and vandalism. They are sunken into the floor and, like the overhead closer, can be adjusted for both closing and

As the door will be used (and abused) by a lot of people, it needs to be well constructed.

Will the door have a self closer or latch mechanism? Getting the right ironmongery is important. It can be very annoying for residents on the ground floor to be kept awake by the constant banging of the door.

Is the door wide enough to take furniture, rubbish bins and other large household objects and can the door be held open when required?

Are the timber sections big enough for the size of the door? Small sections may be more liable to warp.

Where will the door be fixed? If there is a step up to your close, then the door should be set one metre in, past the top step. Unfortunately this

creates an area at the front of the close where people can 'loiter'

Do you want the door to swing both ways? This has certain advantages:
it allows ease of access if people are carrying anything and it avoids the problem of the door banging on the frame. However, it has the disadvantage that you cannot install a security lock on it at a later date.

How much ventilation do you want in the close? Remember that a new close door will significantly reduce ventilation in the close. It may be worthwhile fitting a permanent vent above the entrance door.

Is the timber well seasoned? Red Pine and Douglas Fir are commonly used woods. Hardwood may last longer but is not essential provided

checking actions. They are unlikely to be stolen, but they are more expensive than the overhead closer.

Spring Hinge Closers

These come in pairs: one spring provides the closing action and the other spring provides the checking action. If the door is very heavy, an additional spring may be required.

The 'Dictator'

This is not strictly a door closer, but is very useful in providing a definite checking action to prevent doors banging. The closer is fitted at the head of the door and stops the door on a rubber stopper before it hits the frame. It then closes the door very gently and firmly.

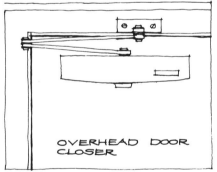

OVERHEAD DOOR CLOSER

THE 'DICTATOR' DOOR CLOSER

OPEN   CLOSED

FLOOR SPRING

HELICAL SPRING HINGE

14.9 DOOR CLOSERS

The closing action of the door may fail because:

The hydraulic oil is squeezed out of the mechanism because people force the door closed. Some closers are designed with by-pass systems which prevent this happening.

The lever arm is pulled down or vandalised and then starts to snag the top of the door. There is a sliding arm closer which avoids this problem.

Wind pressure varies from day to day. An open vent into the close can help equalise pressure differences which may effect performance of the door closer.

Close Windows

Most close windows are **sash and case** *(see window chapter for constructional details: page 81).* If properly maintained they can last a long time. Some windows have decorative glass or small coloured panes, giving a distinctive touch to the close. It is possible to find coloured glass replacement, usually from a stained glass supplier.

Repairing a Close Window

Rotten cills can be replaced without renewing the complete **window case.**

If the bottom stile of the window is slightly loose it can sometimes be strengthened with an angle bracket or timber dowel. If the window is in poor condition, a replacement sash can be made.

If vandalism is a problem and you need a tougher glass, then it is possible to use a

14.10 The Close Window

laminated glass or a tough plastic like *'lexan'.* Both have their drawbacks: the laminated glass may be heavier than the original glass, so the counterweights may need replaced; the plastic gets scratched easily and can be damaged by flames *(e.g. cigarette lighters).*

Close windows can be replaced by pivot windows but take care that the window opening is:

Not too big; some top floor close windows are very high and not suited to pivot type sashes.

Not obstructed by balustrades or stairs which might obstruct the opening of the window.

Close Walls and Ceilings

Close walls are invariably built of brickwork and plastered on both sides. The walls around the staircase are structural in that they support the stair and landings. The walls on either side of the close may only have one wall which is loadbearing, the other wall sometimes stops at ceiling level.

The bottom section of the wall is usually finished in **cement render** or tiled and is termed the dado. The upper section of wall is plastered like the ceiling. Some early closes were never tiled, but much of Glasgow's Victorian tenements are tiled at the ground floor. Some later closes were tiled to the top of the stairs *(a wally close).*

79

What to Look Out For

Always try to check the internal **safe lintols** at the front and back entrances. Check if there are any external cracks to the external door lintol as this can suggest rotten timber lintols on the inside. If there is a wall positioned above the centre line of the close, you may see other small cross beams spanning the close. Check for any sagging here.

Close walls are tall and thin, so they can be unstable. Look out for bulges or bowing of the close wall.

Check that plaster, tile and render finishes are adhering to the brickwork. Old plasterwork can lose its strength and become very weak. Knock the surface - if it sounds hollow **(boss)** the plaster may need to be removed.

Cracks in plasterwork are likely to occur at junctions with other walls, around flat doors and ceilings. Large cracks suggest structural movement. Hairline cracks suggest the plaster may be boss.

The plaster immediately under close skylights is usually set on lathes on a timber frame. Leaking skylights can saturate this plaster and lead to rot in the timber backing.

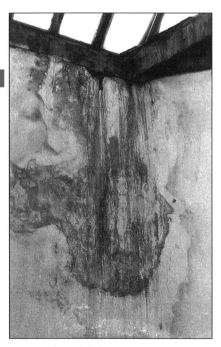

14.11 Leaking skylights are a common problem

Repairing Plasterwork

If the old plaster wall surface is rough and uneven, then it should be hacked off and replastered in two coats of plaster. It is rarely successful to skim over the existing plasterwork since it is difficult to obtain a good 'bond' between the **skim coat** and the old plaster.Cracks should be cut back before filling with plaster.

Cracks at front door entrances are common because of the movement of the timber door frame which is embedded in the brickwork. Cracks in this area are best repaired by stripping the affected plasterwork and fixing galvanised steel mesh over the area before replastering. The mesh provides reinforcement to the plaster.

Separation cracks (the vertical cracks which often appear between the back wall and side walls of the close) appear because side walls are often poorly tied in and the plaster has little to adhere to. Fitting a piece of **galvanised angle mesh** in the corner before replastering can reduce this effect.

Renewing the Dado

Victorian tiles were normally laid on a cement base with very little grouting between the joints. If the close wall has **rising damp**, the moisture in the wall can throw off the tiles. If it is a rendered dado then the render can separate from the close wall.

As dampness is the main cause of tile and render failure, it is wise to check if the close walls are affected by rising damp.

Before injecting a **D.P.C**, the old render should be removed. If the dado is tiled, and the tiles are solid, then they can be left in place. After the D.P.C is injected,

the wall should be left to dry before it is re-rendered using a cement based render.

If new tiles are laid they should be fixed to the render using a cement based adhesive. There is a wide range of tiles available, with special tiles to form a skirting (useful when cleaning the close); and the **dado rail.**

Close Decoration

Decoration is not strictly eligible for a repair grant, although it may be considered essential when it is included in a major repair scheme.

The dado and skirting are normally painted in gloss. New cement render is very difficult to gloss satisfactorily because of the roughness of the render surface. Such render may need additional primer coats or a proprietary filling before the undercoat and gloss is applied. Alternatively a **hardwall plaster** may be applied over the render to give a smooth finish.

Old paintwork may be so encrusted that it needs to be burned off first.

The walls and ceiling above the dado were often whitewashed. When this is painted with an emulsion paint, the old whitewash must be scraped down and the whole wall 'sealed' with a special sealer. If the wall is not sealed, the emulsion may not adhere properly and stains from the old plaster can leach through to the surface.

Cast iron railings should be wire brushed or the old paint burnt off completely before repainting with undercoat and gloss.

The handrail can be sanded down and revarnished, but this should be left to the very end to avoid damage from other trades or dust in the atmosphere.

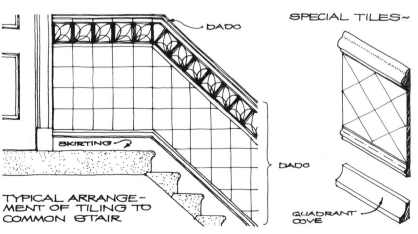

14.12 CLOSE TILEWORK

Windows

The most commonly found window type in Scottish Victorian tenements is the sash and case window. Although the other types are described briefly, we have concentrated on the sash and case window in this chapter.

The Sash and Case Window

Each window usually has an upper and lower **sash** (*these are the sections that move up and down*). Some larger windows may require three sashes. The case is the section which fits into the window opening and incorporates a timber cill. The window sashes are held in place by **counterweights** which are housed in the **pulley box** at either side of the case.

The window would normally be made from Scots pine or an imported redwood.

There have been a number of changes made to the design of sash and case window over the years. In the 18th century, crown glass had to be used. This could only be obtained economically in small sizes and were glazed with 6 or 9 panes. The glazing bars which form these small panes are referred to as astragals (after their original shape). After 1845, a tax on the weight of glass was abolished, and heavier drawn sheet glass was used. This required heavier astragals. Eventually glass production improved to the extent that the Victorians did not require the astragals and it was common to glaze in one or two sheets of glass. This added weight of glass on the top sash led to the strengthening of the joints. This strengthening takes the form of horns on Victorian sashes.

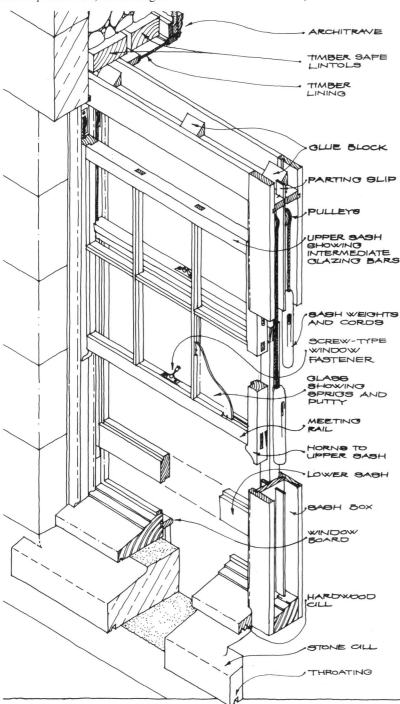

ARCHITRAVE
TIMBER SAFE LINTOLS
TIMBER LINING
GLUE BLOCK
PARTING SLIP
PULLEYS
UPPER SASH SHOWING INTERMEDIATE GLAZING BARS
SASH WEIGHTS AND CORDS
SCREW-TYPE WINDOW FASTENER
GLASS SHOWING SPRIGS AND PUTTY
MEETING RAIL
HORNS TO UPPER SASH
LOWER SASH
SASH BOX
WINDOW BOARD
HARDWOOD CILL
STONE CILL
THROATING

15.1 THE SASH AND CASE WINDOW

What to Look Out For

Some of the window replacements carried out in the 1980s were very poor. Shortcuts were made by not removing the old window frame completely. The linings of the case would be left and the old timber cill cut back. The new window was then slotted into the old case and the gap left at the old cill would be covered over with a facing timber. The lack of a proper cill to throw water to the outside can lead to water penetration and possibly rot to the areas below the window.

Ironmongery

The advantage of the sash and case window is its ease of maintenance and simplicity. The following fittings are commonly found on a Scottish sash and case:

Sash Fastener

These are usually made of brass. There are three main types:

> *The straight arm sash fastener is the commonest. Care has to be taken to ensure the two rails meet up before closing the fastener.*

> *The Fitch Cam fastener operates in a similar manner to the straight arm, but tends to draw the two sashes together. It is more forgiving if the sashes don't meet tightly, but it does not have a return spring so it is more easily damaged.*

The screw type fastener. This is best used if the rails do not meet perfectly. The tightening of the thumbscrew draws the rails together and provides better security. The disadvantage of this type is that the fastener can drop down and snag the top rail of the bottom sash when the window is opened.

Sash Lifts

Simple **hook and ring sash lifts** are designed to ensure the pulling force is kept as near to the centre point of the window as possible. If they are set outwards, increased friction develops making the window more difficult to open.

Original **sash handles** should be retained if possible.

The top sash should be fitted with a **sash eye** so that a sash hook can pull down the top sash.

Pulleys, Weights and Cords

Each sash is attached by sash ropes which feed over axle pulleys into the **pulley box** where they are attached to counterweights. These counterweights may be lead or cast iron, depending on the weight of the window. The weights for the two windows will be different because the top sash must close upwards and the bottom sash must stay closed on the cill.

The method of calculating the weight sizes is shown by the following example:

Top sash weighs 30lbs.

Weights for top sash should weigh 30lb + 2lb = 32lb.

Each weight should weigh 16lb.

Bottom sash weighs 30lbs.

Weights for bottom sash should weigh 30lb - 2lb = 28lb.

Access into the pulley box to retrieve or record the weights is obtained by removing the **pocket pieces** at the base of the **pulley stiles**.

The cord used is traditionally a cotton sash cord but it comes in various grades depending on the window size. Rot-proof nylon cord is now available and should last longer. Cords should never be painted.

Simplex Fittings

These enable the lower sash to open inwards for cleaning and maintenance. Two **butterfly hinges** screwed to the left hand side batten rod,

enable two screws in the sash to slot in to them, forming an inward opening window. However for this to happen the **batten rod** on the other side has to be removed, and the cord on the right hand side detached from the sash. A batten rod screw allows the batten to be easily removed **(some have batten rod hinges as well although these are less essential).** The cord is fitted with a **cord grip** which hooks over a screw in the side of the sash. The cord is retained temporarily by activating the cord clutch, fitted near the sash pulley.

Security

The window can be locked with a **sash screw** which ties together the two sashes at the mid rail *(a simple screw will also suffice).*

A **sash stop** can be fitted which prevents the window from opening more than 100mm. There are two types: *a screw out bolt which is operated by a special key and the Wilkes sash stop which incorporates its own childlock.*

Draughtproofing

One complaint often levelled against the sash and case window is how draughty it is. In fact a well fitted and maintained window will not be draughty and the slightly loose fit of the sashes has the benefit of allowing some natural ventilation which may be advantageous.

New sash windows can be draughtproofed in the factory, but there are still some teething problems as the window is not really designed to incorporate gaskets and weatherstrips.

Mastic

The gap between the stone and the window will be filled with **mastic compound.** Large gaps (over 20mm) are often filled with newspaper first (or nowadays expanding foam) then a **red linseed oil mastic** is trowelled into the gap. If the gap is smaller, then a **polysulphide mastic** can be used. Once again some form of packing will be required first of all. Small gaps can be filled with a **polyethylene backer tube** which compresses to the right shape. New windows can be prefitted with a compressible foam material before they are offered up into the opening.

Mastic should not be taken out level with the cill, if possible a small drip or gap should be retained between timber and stone cills.

Maintenance and Decoration

Because the traditional sash and case window is made from timber, it requires regular painting to prevent moisture saturating the wood and causing wet or dry rot. The most vulnerable parts of the window are the **cill** and **bottom rails** of both sashes. These are exposed to most of the water run-off. Internal condensation on the window can also seep into the bottom rails, causing the putty to detach and speeding up the weathering process.

The windows should be inspected thoroughly before repainting so that any repairs can be attended to first. A screwdriver pushed into the vulnerable areas will soon show if there is any decay.

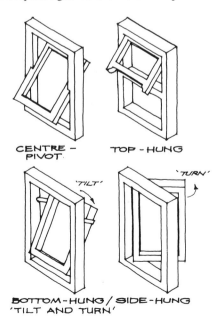

CENTRE - PIVOT.   TOP - HUNG

'TILT'   'TURN'

BOTTOM-HUNG / SIDE-HUNG 'TILT AND TURN'

15.2 MODERN WINDOW TYPES

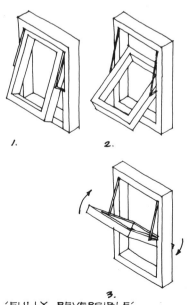

1.   2.

3. 'FULLY REVERSIBLE'

Cills can be replaced without too much difficulty. The bottom section of the case may also need to be spliced in.

The joints of the sash may become loose. Provided the wood is not rotten, the joints can be strengthened by drilling holes and glueing in wooden dowels. Alternatively metal angles can be used. These are equally effective, although more unsightly.

Always check the state of the putty and mastic. Painting over putty which is about to fall out will not lengthen the life of the window. Provided they are instructed, most painters will include a sum for replacing defective putty and mastic as required, or as specified.

New windows should receive a minimum of one primer coat before being glazed, then two undercoats and a gloss when white paint is used. Some manufacturers can supply the windows pre-painted, this ensures good adhesion of the paint. Stains are also used, but they too will require regular maintenance, and remember that the stain has to be carried through to the inside face as well. **Microporous paints** are now on the market. These allow for the natural movement and breathing of the timber and are said to last longer.

Replacement Windows

If the windows are in very poor repair, it may be possible to obtain a repair grant towards the cost of replacing them. Before you agree to anything, check with the Planning Department for advice on what window types are allowed in your area. If the wrong type is fitted without approval *(say a pivot type in a conservation area)*, the Planning Department have the power to insist the pivot window is replaced with a more suitable design.

Removing the old window will cause a lot of disturbance, so make sure the furniture and carpets are well protected from dust. The timber panels or shutters on either side of the window will have to be temporarily removed to allow the complete case to pull out.

It is wise at this stage to check the condition of the timber **safe lintols**. These are very vulnerable to rot outbreaks because they are embedded in the wall. If the window cill above has decayed, water can trickle down behind the panelling and encourage an outbreak of rot to these lintols.

The replacement window should be fixed to the existing wall with rustproof brackets. The timber panelling at the **ingoes** is then replaced.

Pivot Windows

Vertical pivot windows have a hinge at the top and bottom of the frame, horizontal pivots are hinged on both sides. They are pivoted so that a portion of the window goes outside and a portion inside when opened. The junction between frame and sash can be easily draughtproofed. Because tenement windows are generally tall, two pivoted windows and a central **transom** are required. Cheaper substitutes have a fixed bottom pane and a single pivot window.

Tilt and Turn Windows

These are dual action windows. The top of the window can hinge inwards to give top ventilation. If the window is closed and the handle position changed, the window will hinge inwards like a casement. They are often used as fire escape windows because of the ease of access.

Fully Reversible Windows

These are similar to horizontal pivot windows except the hinge mechanism is a metal arm which allows the window to rotate outside the window frame. They can be designed in two sashes without the need for a central transom, so they appear very similar to a sash and case window, except when they open.

Cleaning Windows

All windows should be capable of being cleaned from a safe position. Pivot and fully reversible windows should have a **safety lock** on them once they have been rotated for cleaning. Fixed panes should only be positioned below the opening window, and no more than 300mm deep. Sash and cases are more difficult to clean because the top sash cannot be reversed, only lowered and cleaned by sitting on the cill.

Window Materials

Window frames are built out of four main materials: *timber, aluminium, steel* and *plastic (UPVC)*. Each material has its advantages and disadvantages, briefly summarised here. In the majority of cases, timber is the most appropriate material when replacing windows in

tenement properties. While hardwood cills may be specified for greater durability, most windows need not be hardwood. Therefore unless you can be assured that the wood comes from a renewable source, it is preferable to use a softwood.

Material Advantages and Disadvantages

TIMBER
Advantages

Durability of good timber proven; relatively cheap; unlimited colours; good insulator; renewable and easily repaired; mouldings and shapes possible.

Disadvantages

Needs redecoration every five years; affected by moisture and liable to rot if not maintained; quality of construction and material can vary.

ALUMINIUM
Advantages

May be factory coloured, requiring little painting; quality of material assured; good weatherstripping and tight fitting; relatively rustproof.

Disadvantages

Relatively expensive; difficult to repair if damaged; colours limited; not very 'green'.

STEEL
Advantages

Provided it is galvanised, unaffected by moisture; colours unlimited and easy to change; relatively cheap; thin sections possible; quality assured.

Disadvantages

Needs redecorating every five years (unless factory coated); poor insulator; can twist with age; difficult to repair.

PLASTIC
Advantages

Decoration not required as self coloured; unaffected by moisture; good insulator; special sections easily formed; good for draught seals and double glazing units.

Disadvantages

Untested long term performance in UK although used for 30 years in Germany; limited colours; regular cleaning required; difficult to repair if damaged; can be melted.

Double Glazing

There are basically two systems of double glazing:

> Sealed units which are fitted directly into the window frame.

> Secondary glazing which is fitted as an addition.

Sealed Units

The two sheets of glass are separated by a spacer, to give a gap varying from 4mm to 24mm, and the edges are sealed with a special mastic sealer. This seal is important because it prevents moist air getting into the cavity and condensing. New glazing units should have a 5 year guarantee at least.

This type of glazing is mainly used to reduce heat loss, a 12mm gap being more effective than a 4mm gap. Because of the greater weight of double glazed units, **glazing beads** are used to fix them into the sash. A vented bottom bead is recommended as it allows moisture to drain away from the edge seal. They can also be fitted into sash and case windows, but the frame has to be slightly deeper and the counterweights much heavier. If the window is designed to have astragals, they will have to be thicker and deeper to cope with the glass thickness. The effect can be unpleasant.

Secondary Glazing

If you have existing sash and case windows, a secondary window can be fitted on the inside. It can be designed to hinge inwards or slide, depending on the window configuration. The gap between the two panes is usually much greater (about 100mm). This larger gap is very effective at reducing noise, making this kind of glazing appropriate where traffic is a problem. However if noise is to be reduced the window cannot act as a room ventilator. Separate acoustic wall vents may be needed if there are no chimneys to provide natural ventilation.

Internal Walls, Floors and Ceilings

Internal Walls

The internal walls of a tenement help provide stiffness to the structure, as well as provide restraint to the outer walls and carry the weight of the floors above. They were commonly built from timber studs or brickwork. Walls which carry the weight of other floors are called loadbearing.

Loadbearing Walls

The floor joists of a Victorian tenement are generally carried in one piece from front to rear walls. However, they span between the external walls and the central internal walls. These internal walls are usually the bed recess or corridor walls, and because they carry part of the floor load are, along with the external walls, called loadbearing walls. Under any improvement or repair scheme, care has to be taken to ensure these walls are not removed or their strength and stiffness reduced by the removal of adjacent walls.

When walls have to be removed or altered, it is advisable to consult a structural engineer or architect first. Even small alterations inside may require building warrant approval.

Internal Brick Walls

Most internal brick walls are built from a single leaf of brickwork, 110mm thick, bonded with lime mortar. The lime plaster coating on either side of the wall is built up in two or three coats depending on the straightness of the wall. The overall thickness of the wall is about 150mm. Thicker walls may be used to enclose ducts or chimney flues.

These internal brick walls are normally carried down to their own foundations. The walls in the solum area may be built in a honeycomb fashion to allow through ventilation. However some internal brick walls are built off stiffened timber joists (usually above shops), and movement of the walls is not uncommon. In order to provide stiffness and ease construction, internal door frames were often built between floor and ceiling in a large 'H' section. Sometimes the panel above the door is glazed, but often it is filled with brickwork and the timbers plastered over. This can cause problems if the door frame has to be removed.

Timber battens are sometimes fixed into the walls for two reasons:

To provide a suitable fixing for shelves or fixtures.

To increase the lateral stiffness of the walls (often found on the top floor walls where the lack of self weight allows the walls to flex more).

Timber Stud Walls

Timber stud walls were frequently used in Georgian and early Victorian buildings. The walls were built from vertical timbers known as studs and stiffened with small horizontal sections called dwangs. As plasterboard was not in common use until the 1930s, the background for plaster was made from small timber lathes, nailed close together but with a small gap to provide the plaster with a key. The plaster would be mixed with hair (originally from oxen or goats, cheaper versions mixed with manilla fibre) to give it more tensile strength at these gaps.

Timber stud walls will sound hollow when tapped. Like the brick walls, they

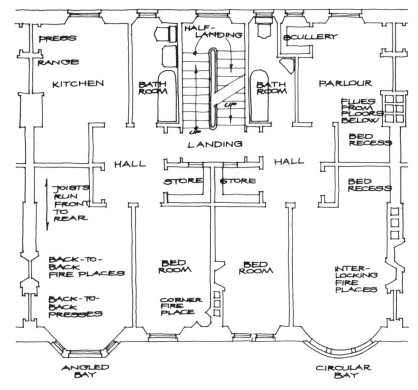

16.1 TYPICAL TENEMENT FLOOR PLAN

What to Look Out For

Poorly bonded brickwork, or very weak lime mortar, may allow the slender walls to become unstable. Brick walls are often poorly bonded into the external walls.

It is inadvisable to remove the plasterwork from both sides of a thin brick wall at the same time, as the plaster may be the only thing holding the bricks together.

Vertical separation cracks can develop and strengthening may be required.

Plasterwork can lose its adhesion to the brickwork. The plaster will sound hollow (or boss), and large sections of it can easily be removed.

Cracks in the plaster may be the result of old movement in the wall. If the backing is sound, the cracks should be cut back and filled with plaster.

can be loadbearing and act in a structural manner. It is dangerous to assume that because a wall is timber stud it can be removed.

In order to provide additional sound proofing, the builder may have filled the space between the timber studs with loose bricks, or even applied a double layer of lathe and plaster.

New timber stud walls are often used in rehabilitation schemes to form new bathrooms or kitchen compartments. The timber frame is constructed in the same manner as old stud walls, but plasterboard is used to sheet both sides. This can be self-finished or given a skim coat of plaster. Double sheeting the walls with different thicknesses can provide additional soundproofing.

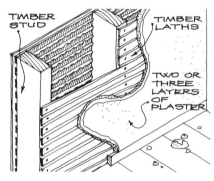

16.2 TIMBER STUD WALL

Wall Linings

External walls are usually finished on the inside face with a wall lining of lath and plaster. This will sound hollow when tapped (if it sounds solid then the plaster may be applied directly to the brick or stone wall). The lathes for the plaster are fixed to vertical timber straps which are fixed to the wall using timber dooks embedded into the stonework joints. The walls are referred to as strapped and lined walls.

The modern equivalent of external lining will use plasterboard. Insulation will be either glass fibre matting fitted between the timber straps, or a composite insulation board using polystyrene or other foam insulants. Whenever insulation is added to the external or close walls, a vapour barrier may be required.

Skirtings

The gap between wall plaster and floor is finished with a timber skirting. The skirting is fixed to timber grounds which are fixed to the walls. Large moulded skirtings are often made from two (or

more) pieces. Gaps which open up under the skirting are sometimes filled with a quarter round section of timber called a mouse moulding.

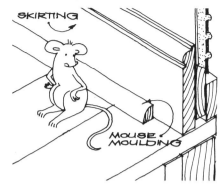

16.3 MOUSE MOULDINGS

What to Look Out For

As the external stone wall does not have a cavity, its internal face may become damp on occasions. This slight dampness, combined with the right temperature and humidity, can allow rot to develop in the timbers behind the plasterwork.

Like the brick walls, plaster on lathes can lose its key and become boss. This will often happen if water has soaked the plaster, but movement, drying out of plaster, and rot can all affect plaster adhesion.

Timber lathes are very vulnerable to woodworm and rot.

Some walls may sound like brick, but are built from lathe and plaster laid in very thick coats.

Floor Construction

Internal floors were built using timber floor joists which, on the upper floors, were laid in one piece from the front to rear walls, but spanned between external and internal loadbearing walls. On the ground floor, smaller floor joists were usually built onto *sleeper* **or** *dwarf* **walls.**

The main joists would be stiffened to prevent flexing by adding **herringbone struts** or solid timber **dwangs** between each joist.

Soundproofing was provided in a number of different ways depending on location and date of construction *(see soundproofing floors on page 87).*

The floorboards were usually red pine and **tongued and grooved (t&g)**. The boards could be **secret nailed** by nailing through the tongue at an angle, but this was not always done. It is usually possible to identify the joist positions from the nail holes in the boards. Lifting boards are usually provided above electric lighting drops and at areas where maintenance may be required.

In rehabilitation work, **chipboard flooring** is frequently used. This comes in sheets and is also tongued and grooved although the joints should be glued in addition to nailing. Different grades and thicknesses are available and **water resistant chipboard** floors should always be used in kitchen and bathroom areas. However, unless special service ducts are built in, lifting a chipboard floor for maintenance is likely to damage the floor.

Stone hearths and tiled entrances were built in a similar manner to close landings. A stone slab was laid on a compacted layer of **deafening** which rested on a timber base laid over timber (and sometimes steel) joists.

Levelling Floors

There are two main methods:

The floorboards are removed and set aside. New levelling joists are fixed to the side of the existing joists and the floorboards replaced.

The existing floor is retained and a new floating floor on battens is laid over it. The battens are packed to take account of the level changes. This will inevitably mean that a step is required at the front door.

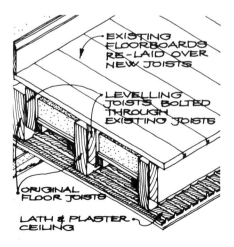

16.4 LEVELLING TIMBER FLOORS

What to Look Out For

Floors off level. Localised subsidence may cause floors to be off level. This is easily checked with a spirit level or a marble. The precise amount of fall can be calculated by using a **water level.** Provided there is no further subsidence, it may be possible to level the floor.

Movement in floors. This may suggest **dry** or **wet rot.** Floor joists which are built into the outer walls are vulnerable. Wet deafening provides an ideal basis for the retention of moisture and spread of rot. Identifying rot in the floor space is nearly impossible unless the floor is opened up.

Squeaking floorboards. If floorboards are badly fitted or gaps are left between board and joist,

squeaks can develop. The loose boards should be screwed down or lifted and replaced.

Replacing floors. If the floor is being completely replaced then a **flooring grade chipboard** may be used. However this may require additional struts between joists at the chipboard joint positions.

Notching joists. This is often carried out to allow electric cables and pipework to pass between the joists. Ideally cables should be taken through holes drilled in the centre of the joists, as large notches can seriously reduce the strength of the joist. **Cover plates** should be fitted over notches to prevent damage from accidental nailing from above.

Installation of a secondary floor layer, known as a **floating floor**. Battens are laid on sound insulating material which is laid over the existing floor, then boarded over. The sound insulating material is turned up at the edges and provides an **isolating membrane** between the wall skirting and the floor.

Special chipboard flooring which has a combined lead foil and felt backing. This is laid directly over the existing floor.

Carpet underlay can reduce impact noise. If it is glued to the floor it may be treated as a permanent fixture.

Deafening can be replaced in small areas with a mixture of sand and finishing plaster. In larger areas small limestone chippings (*trade name Quietex*) can be used.

Double sheeting the ceiling with two layers of plasterboard of different thicknesses.

Forming a separate suspended ceiling under the existing ceiling. The new ceiling should be as isolated as possible from the floor above, thus reducing impact noise.

Some Housing Associations now remove all the old deafening and replace it with a suspended ceiling, and a new floating floor above.

Walls between neighbouring properties present a more difficult problem. If the wall is relatively thin then it may be best to build a separate stud wall alongside it. If possible this should be isolated from the party wall so that the impact sound is reduced.

Soundproofing Floors

How Sound Travels

Sound usually originates in one of two ways:

As a disturbance in the air, such as the human voice or music. This is referred to as airborne sound.

As a blow or movement to some part of the building structure. This may originate as footsteps, doors banging etc. This is referred to as impact noise.

Both types of sound can travel to the neighbouring premises by both means - through the structure and through the air. In the majority of cases within a tenement the problem is one of vibration as sound travels through the structure.

Traditional Soundproofing

Soundproofing in the floors of older tenement properties usually consists of a layer of ashes about 70mm deep set on top of rough boards (**deafening boards**) between the joists. The material is commonly referred to as **deafening.**

In Edinburgh, the construction is slightly different as the ceiling level is often suspended on hangers and branders, creating a larger, and more soundproof, void between deafening and ceiling. Thick plaster coatings and cornices helped to fill gaps between floors and walls.

Sound Testing

Traditional soundproofing may not be up to today's standards, and repair work to

floors can reduce the effectiveness of the original soundproofing qualities. Most local authorities now insist that **sound tests** are carried out in houses before any rehabilitation work starts, so that the reduction levels of the existing and new floors can be assessed. In major rehabilitation works, the sound reduction qualities need to be improved to meet the Building Regulation standards. In minor repair works, the quality of sound reduction can stay the same as the original, provided an initial sound test is taken.

Soundproofing Remedies

There are a variety of options being used to soundproof floors, they include:

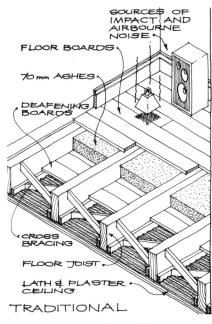

16.5 SOUNDPROOFING TO FLOORS

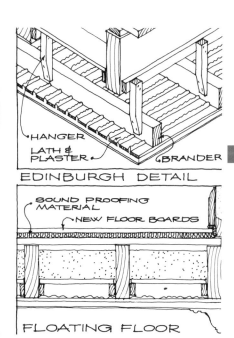

Ceilings

The majority of ceilings, up until the 1930s, were formed using **lath and plaster** *(see internal wall construction on page 85)*. In Glasgow, the lathes would be nailed directly to the underside of the timber floor joists. In Edinburgh they would prepare a separate framework of **branders** hung from the joists by **struts**.

Cornices

These are found in nearly all Victorian tenements and are rarely found in modern construction. The plaster cornices were usually made on site, after the lathes had been formed to shape.

Additional decorative features were sometimes prefabricated and fixed once the backing cornice was built up.

Broken or damaged cornices can be repaired by matching in prefabricated **fibrous plaster** sections which can be

16.6 Dampness weakens the plaster key and can lead to ceiling collapse

made in nearly any shape. The cost of replacing decorative cornicing is high and will not be covered by a repair grant. However if your building is **listed**, or in a **conservation area**, additional grant may be available for the work.

Repairing Old Ceilings

Cracks in ceilings may be the result of movement in the building fabric, rather than the plaster. If the plasterwork is adhering on either side of the crack, it should be possible to repair. The crack should be cut back before filling.

Small areas of plasterwork can sometimes be patched in, provided the backing is firm. Plasterboard can be used to build out to the thickness of the original coat, but invariably small hairline cracks will appear around the patched area.

Renewing Ceilings

The removal of any ceiling will create a lot of dust and plaster debris. It is essential that everything is removed from the room or completely protected *(polythene sheet over carpets should be taped at all joints)*. Once the plaster is down and the lathes and nails removed, plasterboard can be fixed to the underside of the joists. If the surface is very uneven, timber battens called **branders** can be fixed across the joists to form a better fixing for the plasterboard. Indeed some builders may suggest **brandering** the ceiling without dropping the old plaster; this does work but prevents a proper inspection of the joists.

It is still good practice to plaster over the plasterboard with two coats of plaster. The first coat is called a **browning coat** (or **bonding coat** if there are thicknesses to be built up) and the finish is called a **skim coat**. The joints between the sheets and between ceiling and walls should be joined with a **scrim tape** before plastering. If this is not done, small cracks can develop at these junctions. **Dry lining** may be used instead of plastering. In this case the plasterboard has a special joint which is filled with a paper tape and plaster. The surface is then treated with a proprietary sealing coat before being decorated.

What to Look Out For

Plaster ceilings may appear perfectly sound. However, if they are prodded with a stick (broom handle) any movement is likely to mean the plaster has detached itself from the lathes and may detach itself without warning.

Staining on a ceiling may suggest previous water penetration. This may have led to rot in the joists above, or simply caused the plaster to become too heavy to retain its keying action with the lathes.

Chemical staining can occur (chemicals leaching out of treated joists etc). Old plasterwork can be sealed with a sealer before decorating, otherwise stains can continue to leach through.

Plasterwork should always be allowed to dry out thoroughly before decorating. If ventilation is restricted, mould growth can form very rapidly on the surface.

If cornices are removed, the sound penetration between flats may increase as the joint between wall and ceiling is a very vulnerable area.

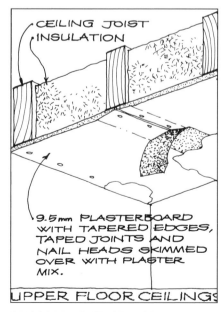

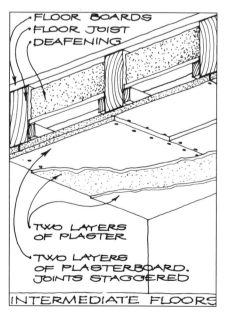

ILLUS 16.7 RENEWED CEILINGS

Plumbing

This chapter deals with the common elements of tenement plumbing systems, rather than the plumbing arrangements of an individual house. Little has changed in the general principles of plumbing since about 1860 although new materials and boiler systems have greatly expanded the level of plumbing services.

In any house there are three distinct elements to the plumbing system:

The water supply which provides fresh water for drinking and cooking and also feeds the hot water supply.

The waste system disposes of waste and water from sinks, baths, washing machines and WCs.

The surface water system disposes of rainwater via rhones, downpipes and gulleys.

The Water Supply

The water main runs under the pavement. At the entrance of each close, a branch is taken off and a toby installed. The **toby** is simply a valve which allows the water supply to be closed off to the complete tenement. A **water key** inserted into the toby hooks over the valve and given a quarter turn should close off the water.

The supply pipe is usually buried under the floor of the close, and originally would have gone into the backcourt to supply water to the boiler in the wash house. Sometimes two pipes are taken off the main supply and re-enter the building at the kitchens on either side of the close. These **rising mains** carry on up through each flat, supplying water to the kitchen sinks, and eventually supplying water to the cold water storage tanks in the loft.

Sometimes there is a single storage tank supplying water to the flats below, alternatively there may be a number of smaller tanks **breached** together. These tanks, like the complete plumbing supply, were usually made from lead and set into a timber coffer. The supply coming into the tank is controlled by a **ballvalve.** If this fails then the water must be able to escape to the outside otherwise the tank will overflow. Warning or **overflow pipes** can be fitted at the top of the tank or in its base. The base outlet is fitted with a removable upright pipe which comes above water level but can be removed to allow the tank to be drained and cleaned. Overflows may extend out at eaves level or be carried down to a convenient position at ground level.

Two separate feeds are taken off the storage tanks:

A common feed pipe which carries cold water to the toilet cisterns, baths and basins.

An individual cold supply to feed a hot water cylinder in each flat.

Stopvalves should be fitted to the supply and feed sides of the pipework, and to the drinking water supply to the sink units.

Although lead pipe was extensively used in the past, modern water supply systems can be made from plastic, copper or stainless steel.

Hot Water System

Most modern central heating systems will use a boiler which heats hot water and room heaters. However many houses still have independent hot water systems. The water is heated in the hot water cylinder by either an electric immerser or an external boiler.

Electric Immersers

Electric immersers come in a variety of types:

Top mounted: a large heating element is fed into the cylinder at the top. A thermostat controls the water temperature.

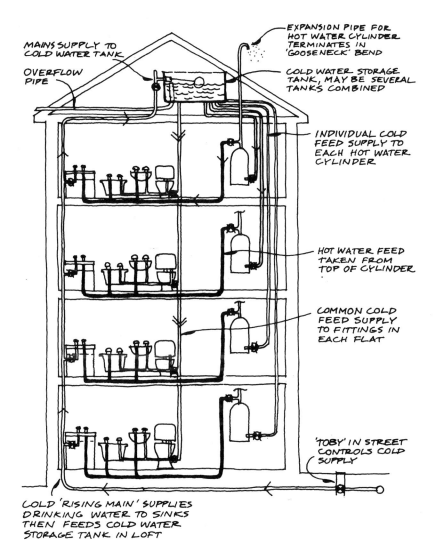

MAINS SUPPLY TO COLD WATER TANK

OVERFLOW PIPE

EXPANSION PIPE FOR HOT WATER CYLINDER TERMINATES IN 'GOOSENECK' BEND

COLD WATER STORAGE TANK, MAY BE SEVERAL TANKS COMBINED

INDIVIDUAL COLD FEED SUPPLY TO EACH HOT WATER CYLINDER

HOT WATER FEED TAKEN FROM TOP OF CYLINDER

COMMON COLD FEED SUPPLY TO FITTINGS IN EACH FLAT

'TOBY' IN STREET CONTROLS COLD SUPPLY

COLD 'RISING MAIN' SUPPLIES DRINKING WATER TO SINKS THEN FEEDS COLD WATER STORAGE TANK IN LOFT

17.1 THE WATER SUPPLY

Circulator immerser: a heating element is fitted at the base of the cylinder and an internal tube is fitted to it. This allows the hot water to circulate to the top of the tank, thus allowing a basinfull of water to be heated quickly.

Dual immerser: top and bottom immersers can be fitted, the top one connected to the off-peak supply and the bottom one to the day-time rate.

These immersers are usually rated at about three Kilowatts and have to be connected to the mains supply with their own fuse and isolating switch.

Back Boilers

The water in the cylinder can also be heated by a **back boiler.** Pipework from the top side of the tank is taken to a small boiler at the back of the fireplace; the heated water is then fed back into the base of the cylinder by a return pipe. The cylinder should be above the level of the back boiler so that air locks do not build up in the pipework. This system is known as a **direct system.** Sometimes there is enough hot water to heat one radiator, but they cannot be used to heat all the radiators, for that an **indirect system** is required.

As the hot water in the cylinder heats up it will expand. The tank has to be vented to the air through an **expansion** pipe. It usually appears as a small looped pipe (a **gooseneck**) on the roof above the level of the cold water cylinders. *Remember: it is*

the height of the cold water storage tank above the hot water cylinder which ensures there is sufficient water pressure in the hot and cold supplies.

Insulation of Pipes and Tanks

When water freezes in a pipe, it expands with a force so great that it can cause the pipe to burst. The leak may only be discovered when the ice melts. All pipework fixed under ground floors, in lofts or adjacent to external walls should be well insulated either with a double thickness of hair felt or foam insulation.

Cold water tanks in the loft space should be insulated with at least 75mm of insulation on top and sides, but not under the tank. All hot water cylinders should be well insulated. Modern cylinders are lagged with foam insulation in the factory and are a lot more efficient to run than cylinders with loose quilt insulation.

The Waste System

All waste and soil must go through a waterseal (trap) before entering the main drainage system. The trap prevents smells from sewers entering the house. The water in the wc bowl is the trapped water, and there is a trap under each sink, wash basin and bath.

Scotland can lay claim to being the birthplace of the vented trap. It was W.P. Buchan who, in 1875, saw the benefits of venting traps on the house side, thus preventing siphonage and loss of the trap seal (his design had other self cleaning features as well). Although the 'Buchan trap' is sometimes referred to as a 'Buckin' trap, it generally functioned well and can still be found in use today.

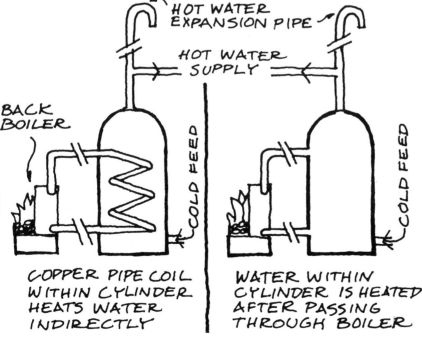

INDIRECT SYSTEM   DIRECT SYSTEM

17.2 INDIRECT SYSTEM, DIRECT SYSTEM

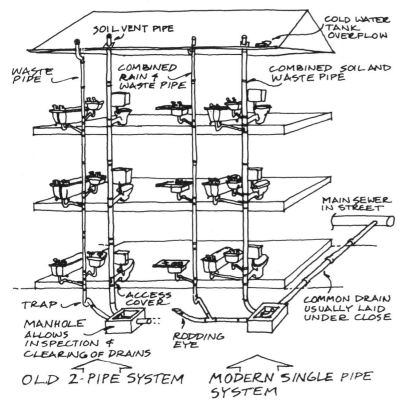

Labels in illustration:
SOIL VENT PIPE
COLD WATER TANK OVERFLOW
WASTE PIPE
COMBINED RAIN & WASTE PIPE
COMBINED SOIL AND WASTE PIPE
MAIN SEWER IN STREET
TRAP
ACCESS COVER
MANHOLE ALLOWS INSPECTION & CLEARING OF DRAINS
RODDING EYE
COMMON DRAIN USUALLY LAID UNDER CLOSE
OLD 2-PIPE SYSTEM
MODERN SINGLE PIPE SYSTEM

ILLUS 17.3 THE WASTE SYSTEM

There are two different systems for disposing of soil and waste:

Two Pipe System

Older buildings will work on the two pipe system where one pipe takes the soil from the WCs and another pipe collects waste from the bath, wash basin and sink. This waste pipe would also be trapped before it entered the sewer system.

Single Pipe System

The modern system is to use a single pipe (or **stack**) into which all waste flows. However it is still common practice to connect kitchen waste pipes to rainwater down-pipes. In such cases the pipe should be trapped before it enters the drainage system.

All waste pipes drain into the sewers, usually via a **manhole** which allows access to the drains for clearing blockages.

The top end of the waste and soil stack is ventilated above eaves height. Without this ventilation, water seals at fittings could be sucked out through siphonage, letting smells into the house. A **wire balloon** at the pipe outlet prevents birds from nestbuilding and blocking the vent.

Waste pipes can be cast iron or plastic. Both should have access covers in the pipe near bends to allow the system to be cleared.

Fire Stopping Pipes

Sometimes soil and waste stacks are kept inside the building. This can reduce the risk of frost and the need for repainting the pipes. When such pipes pass through the floor they should be fire stopped. This prevents the spread of smoke or flame from one flat to the next. Cast iron stacks are relatively fireproof, but any gap between pipe and floor needs to be sealed with cement or **vermiculite.** Plastic pipes can melt, so they should be fitted with an **intumescent fire collar** which expands and closes the pipe if there is a fire.

Drains and Sewers

The individual stacks at the back of the tenement will be fed into an underground drain which remains the responsibility of the owners until it connects into the sewer in the street. The drains can be made of plastic or clay pipe. **Manholes, intercepting traps** and **rodding eyes** are often found at junctions and bends to facilitate cleaning the drains.

The pipework must be laid on a gravel base and to the correct fall. Pipework must also be jointed smoothly if blockages are to be prevented.

Sewers are the responsibility of the Regional Council. Sometimes your drain will go into a branch drain which is shared by a number of closes, before it enters the main sewer. These drains are usually the responsibility of the Regional Council as well.

Overflows

As a safeguard against flooding, all wc cisterns and water storage systems must have overflow pipes which discharge to the outside. Baths, sinks and basins also have overflows, but they can discharge directly into the waste pipe system. Continuously running overflows can be reported to the Regional Council as they have powers under the Water By-laws to stop the waste of water.

What to Look Out For

Broken soil and waste stacks. Cast iron is generally very robust, but it can crack or rust internally. The area next to the wall is most vulnerable because it is hardest to repaint. Watch out for staining on the stonework near the pipe. Some pipes can look in good condition, but are so badly choked with rust residue that they are totally blocked.

Blocked Drains. If the manhole or **disconnecting trap** starts to overflow, there is likely to be a blockage between this point and the sewer. If **rodding** does not clear the blockage, special drain cleaning apparatus may need to be used (jet washing and spring coils). If the blockage is still solid then opening up the drain may be the only solution.

17.4 RAT IN A MANHOLE

Rats. If you have seen rats in the backcourt, they may be nesting in the drains. One vulnerable place is the shelf area in the manhole. Rat poison left in the manhole may help, otherwise contact your Environmental Health Department.

Gutters Valleys and Downpipes

Gutters (*rhones*), are designed to direct water from the roof into the downpipes. They need to be large enough to cope with the size and pitch of the roof. Gutters will overflow if they are not big enough, or if there are too few outlets into downpipes.

Front Gutters

In Victorian times these were usually made from cast iron and moulded in shape. The abbreviation 'O.G.' in Scotland refers to an Ornamental Gutter.

What to Look Out For

Inspections from the ground are best made after a fall of rain, as evidence of leaking joints often shows up on the stonework below.

Corrosion: cast iron gutters often corrode from the inside face, so it is impossible to assess this from the ground.

Junctions with adjoining gutters. If the gutters are not the same size or shape, the joint is liable to leak.

Signs of gutter blockages or overflowing. This may have led to dampness in the rafter ends leading to rot outbreaks.

The gutter rests on a stone corbel or cornice. Gutter joints are bedded in putty and joined together with nuts and bolts. Some of these gutters have lasted for nearly 100 years, so cast iron has proven to be a long lasting material and can still be obtained for replacement or repair work. Gutters should be cleaned out at frequent intervals (every three years) otherwise sediment builds up which reduces the flow capacity of the gutter.

Gutter Types

In addition to **cast iron** gutters, **glass fibre** and **aluminium** gutters can be used. Their qualities may be summarised:

Glass Fibre: can be made to match most shapes, curves and colours. It is very light and depending on thickness, has a tendency to flex. **Storm clips** should be used to ensure gutter is not bent over by snow loads. The life expectancy of glass fibre may be limited to about twenty years, depending on quality of fabrication.

Light Gauge Aluminium: this is thin enough to be extruded in one length from a forming machine brought onto site. This type of gutter is too lightweight for tenement construction.

Cast Aluminium: this is supplied in similar sections as cast iron, and is jointed with a flexible mastic. It can be

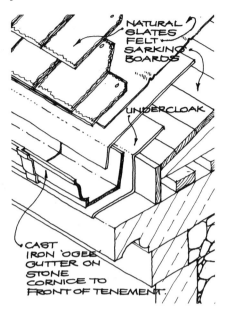

17.5 FRONT CAST IRON GUTTER

supplied in a range of pretreated colours, but is not available in curved sections (for bay windows).

Heavy Gauge Pressed Aluminium: this can be pressed into a moulded shape and is also available in a range of prefinished colours. The joints are sealed with mastic and an internal cover plate. The material can be welded to form special shapes and connections and is favoured because of this flexibility.

Front Gutter Replacement

It is now good practice to ensure front gutters are laid on top of a lead or *nuralite* **undercloak.** This undercloak is set over the corbel, under the gutter and up behind it to a point well above the top level of the gutter. This prevents the water from an overflowing gutter from penetrating into the wallhead and causing rot in the timbers.

Gutters should provide a shallow fall to the outlet (1:50) and should be water tested for flow and leaks before the scaffolding is dismantled. A wire balloon over the gutter outlet should prevent debris from blocking it up.

It is a simple matter to carry out a water test. First of all, seal the outlets with a stopper then fill the gutter to the top. Check to ensure the fall towards the

outlet is satisfactory. Leave the water sitting for about ten minutes, then check all the joint connections for leaks.

Lead Cornice Gutters

Sometimes a gutter is made by forming a channel in the top of a large **stone cornice**, then waterproofing it with lead or asphalt to form a gutter. They are rarely satisfactory because the lead or asphalt is not able to move independently of the stone.

Renewal in lead is possible, but you will need to ensure the correct thickness is used, and allow for an experienced plumber who can do leadburning. If lead is being laid in long lengths, special expansion joints (*T-Pren*) can be used to allow the lead to expand and contract without buckling.

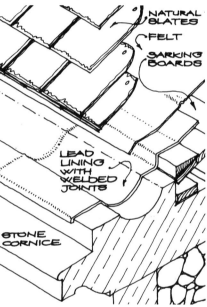

17.6 FRONT LEAD CORNICE GUTTER

Rear Gutters

Rear gutters are usually made from cast iron which is half-round in section. The gutter is supported from galvanised metal brackets fixed to the timber sarking of the roof. As fixing a gutter often requires scaffolding, it can be as cheap to replace the gutter as repair it.

A New Rear Gutter

Cast Iron: new half-round gutters are still available, and have a reasonable lifespan if painted regularly. They also withstand a lot of knocks, so gutters which are accessible from ground level may be best in cast iron. The joints are sealed with a flexible mastic. Although putty is still favoured by many plumbers for cast iron jointing, it has a tendency to dry and crack with age.

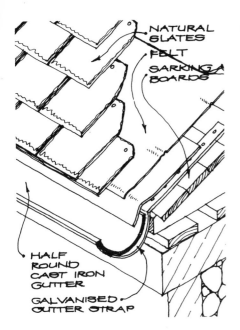

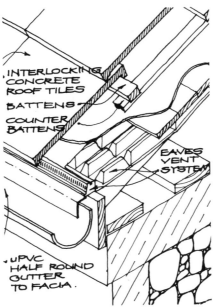

17.7 REAR CAST IRON GUTTER

17.8 REAR PVC GUTTER

PVC gutters are now commonly used as they are supposed to be easier to maintain. However PVC does deteriorate with age so don't expect it to last forever. Deepflow sections will cope with a greater capacity of water than a half round section of similar width. The gutters are joined with proprietary neoprene gaskets which allow for some

thermal movement. They are hung from adjustable **gutter straps** fixed to the sarking (under the felt).

If the gutter is carried past a chimneyhead then surface mounted brackets will be needed (*these are often left out and the gutter then sags under the chimneyhead*). Each bracket should be no more than 600mm apart. The recommended fall for a PVC gutter is about 1:50.

Parapet Gutters

Also known as **secret** or **hidden gutters** because they cannot be seen from ground level. These gutters are usually formed in lead and are stepped down at each joint of the lead (to cope with thermal

expansion). The outlet should be formed in a **sump** (or **catch pit**) which is designed to increase the flow at the outlet thus preventing blockages. An **overflow pipe** which extends through the parapet is another essential, as a blocked downpipe could allow the parapet gutter to overflow into the house below. Timber catwalks or snowboards on top of the lead protect it and allow the snow to thaw without obstructing the downpipe.

Renewing a Parapet Gutter

The majority of repairs to parapet gutters will initially require rotworks to be carried out to the timbers underneath. As this work may take some time, it is essential to provide some form of watertight temporary protection during the contract. An independent 'tent' structure over the whole work is the best method.

17.9 SAGGING GUTTER AT CHIMNEYHEAD

What to Look Out For

Parapet gutters are vulnerable to water penetration, so always **check for rot** under the timber supports.

Check the leadwork for cracks or movement. If the lead has been covered over with bituminous material it is unlikely to last. If possible, try to inspect the jointing between outlet and downpipe - this is a vulnerable area.

If overflow pipes are too small or choked, then water could flood into the building in the event of a downpour.

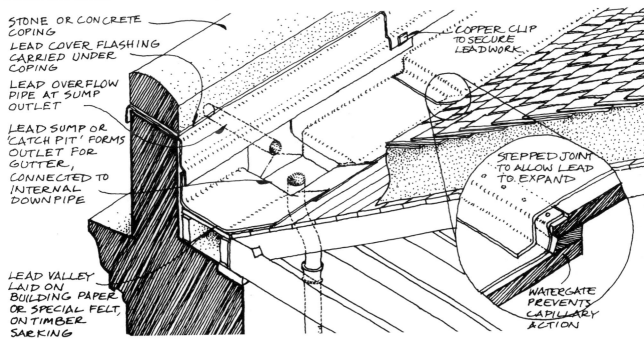

17.10 PARAPET GUTTER

93

Once the old lead is removed and the rotworks attended to, the base boards are renewed allowing for 50mm steps at every joint in the lead. **Building paper** or a **geotextile felt** is used as an underlay to separate the lead from the baseboard. The lead will lap up the side of the parapet and be covered with a lead **cover flashing**. Fixings to the timber sarking under the tiles should be copper.

Overflows need to be fitted at the most vulnerable points. They should be at least 75mm diameter to allow them to take any surge of water from the roof if the downpipe blocks.

It is possible to avoid the need for a parapet gutter by boarding over the area and making it part of the roof. The new boarding is covered in lead and a new aluminium or cast iron gutter is fitted to the external face of the parapet. But be careful, as this option may not be acceptable to the Planning Department.

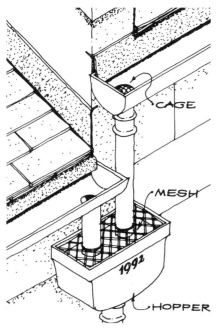

17.11 DOWNPIPE HOPPER

What to Look Out For

Look out for signs of damp or stained stonework. Moss and plants growing from the pipe or stone is a sure sign of a broken pipe. Because cast iron pipes are fixed close to the wall, the hidden area of the pipe is difficult to paint and therefore more liable to rust. Cast iron pipes will also rust from the inside and this can eventually lead to blockage of the pipes.

Downpipes

Rainwater Pipes: pipes which carry rainwater only are known as rainwater pipes (**rhone pipes**), and are usually about 75-100mm in diameter.

Waste Stacks: pipes which carry water from kitchen sinks, basins or baths are termed waste stacks. It is quite common to have sink wastes sharing the rainwater stack.

Soil Stacks: pipes which carry soil waste from the w.c. are termed soil stacks. They are usually about 150mm in diameter and may take waste from the bath and basin as well. However they must not take rainwater.

Most old downpipes are made from cast iron, secured to the wall with metal brackets called **holderbats**. In Scotland it is traditional to take the gutter outlet directly into the downpipe. The connecting pipe is called an **offset**. If two outlets feed into the same pipe, a **hopper** may be used. The advantage of the hopper is that if the pipe blocks it will overflow at the hopper and not at the gutter where it could cause more damage. However hoppers are notorious rest homes for birds' nests so make sure the open top is meshed over.

Repair and Renewal

When blockages appear they can sometimes be cleared by rodding into the pipe at **access points**. These access points should be sited at ground level, or at critical bends and junctions. If none exist the pipe may have to be broken into. New sections can be fitted back into a cast iron downpipe using *timesaver* joints.

If scaffolding is being used then it may be worthwhile renewing the pipes in plastic. It is best to keep the bottom section in cast iron as it is less likely to be vandalised or accidentally damaged.

Plastic pipes should be fixed more frequently than cast iron, as the pipes can flex and may even dislodge from the gutter outlet. All the joints should be sealed with a **rubber sealing gasket**. If the pipes are internal and hidden behind ducts, it is essential that they are completely watertight. Cast iron downpipes are jointed by backfilling with a rope gasket, then **caulking** the gap with molten lead.

If the downpipe takes kitchen waste pipes then these connections should be renewed at the same time, preferably right back to the sink grating.

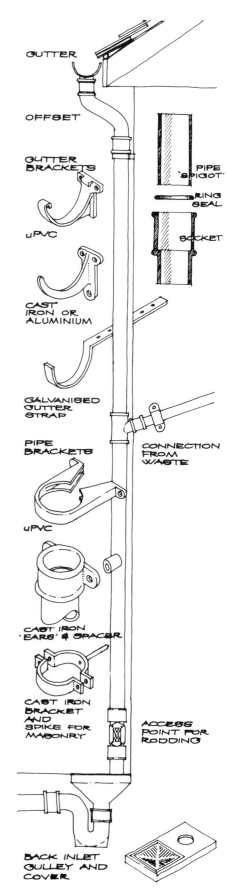

17.12 DOWNPIPE DETAILS

Dampness and Decay

There are four main causes of dampness in tenements: condensation, rising damp, penetrating damp (from walls and roof), and dampness from plumbing defects.

All forms of dampness will eventually lead to decay of timberwork, often leading to dry or wet rot. In this chapter we deal in most detail with rising dampness because it is more prevalent in older housing stock.

Condensation

Condensation has not been a serious problem in old Victorian tenements for several reasons: ventilation is generally good (sometimes excessive); the structure is reasonably insulated; the materials used allow the walls to 'breathe'. However, with the advent of rehabilitation, draught stripping and insulation has reduced the level of ventilation and condensation can occur.

Causes of Condensation

There are three main elements in the building which affect the likelihood of condensation. They are:

The building itself - the thermal properties of roof, walls, floor and type of construction.

18.1 CONDENSATION

Ventilation - both natural and mechanical, concerned with the removal of excess water vapour at source.

Heating - level of heat and frequency of use.

In dwellings where these three elements are correctly balanced, the risk of condensation is generally low. When it does occur the moisture stimulates the growth of **mould spores** (unlike rising damp).

Rising Dampness

Rising dampness results from a capillary flow of water from the ground. To avoid this moisture rising, a **D.P.C (Damp Proof Course)** should be built into the wall slightly above ground level.

Many old tenements still have D.P.Cs made from slate in the outer walls. However it was not common practice to insert a D.P.C into the internal walls, instead reliance was placed on ventilating the **solum** space below the floor so that these walls could dry out.

Surveying Rising Damp

Many specialist D.P.C firms will provide a survey and estimate for you. However, their surveys can vary greatly. Some firms will assume there is rising damp everywhere in order to obtain additional work or simply to 'play safe'. They are not always interested in finding the cause of the problem.

In some cases rising damp is easily observed showing a **tide mark** stain at about three feet from the floor and the plaster below damp. In other cases it can be quite difficult to distinguish rising damp from other causes of damp.

The *Protimeter moisture meter* works on the principal of electrical resistance of the material. It will tell you if the plasterwork or surface of the wall is damp, but it will not tell you why. It may be that the plasterwork is damp but the wall behind is dry (long probe attachments can be fitted to the meter to allow the testing of the wall inside).

The plasterwork may have turned **hygroscopic** due to the presence of chloride and nitrate salts in the ground. Certain salts in soot can also seep through chimney breasts and cause the plasterwork to become hygroscopic.

If the wall plaster contains a high presence of nitrates and chlorides, then it is likely that rising damp is the problem, as these salts are being sucked up from the ground below. A sample of the plasterwork can be sent to Protimeter's laboratory for testing.

A more accurate way of ascertaining the degree of rising damp is to measure the moisture content within the wall.

What to Look Out For

In a tenement the most vulnerable area is usually the close wall as it has the least insulation. The addition of close doors can increase the temperature in the close, and thus reduce condensation risk. However, the close walls should really be insulated on the inside face.

When the loft space is insulated, the houses below will be warmer, but the loft will be much colder. If the loft is not well ventilated, condensation can occur on the underside of the roof.

Large single glazed windows can be very cold and moisture vapour will inevitably condense on the panes. The moisture produced can run down the panes and penetrate the timber cill leading to rot in the bottom rail.

Internal WCs and bathrooms require extractor fans. These are usually connected to the light switch to give a 20 minute 'run on' after the light has been switched off. If this is not done, excess moisture vapour can be added to the atmosphere of the other rooms, increasing the likelihood of condensation. Kitchens are also likely to produce a lot of moisture and extractor fans linked to humidistat switches can help remove excess moisture vapour before it condenses in other colder rooms.

This can be done with a *Speedy carbide meter*, with samples taken from inside the wall at different heights. A moisture profile showing the degree of moisture in the wall is then used to show the likelihood of rising damp.

Other Causes of Rising Dampness

In some cases, the cause of the rising dampness may not be simply from a missing or defective D.P.C. Check the following:

Bridged D.P.Cs: Always check to see if the ground level outside has risen above the D.P.C.

High Water Table: The property may be low lying and the water table relatively high. In such cases the wall footings may be saturated with water, making the installation of an effective D.P.C. very difficult. Ground drainage may help remedy this problem.

Broken D.P.Cs: Structural cracking or subsidence in a wall can fracture a physical D.P.C.

Retaining Walls: Some external and mutual walls may be retaining earth, yet not have a satisfactory vertical damp proof course. In some cases this can only be remedied by inserting an impervious barrier on the other side of the wall.

Burst Pipes: A burst rising main under the ground can remain undetected for years. The surrounding walls may become saturated.

Installing a D.P.C.

If the whole house is being treated you will probably have to move out for the duration of the works. Repair grants are normally available for this work although they do not cover redecoration.

The flooring timbers may also require to be removed and replaced; solum space **tanked** and lowered; and **dwarf walls** rebuilt.

The two main types of installation in use today are:

Physical Insertion of a D.P.C. A deep cut is made in a section of the wall in 600mm lengths. The new membrane is inserted into the cut and the remaining gap fully grouted with an epoxy mortar. The process is not used in walls over 220mm thick.

Chemical Injection of a D.P.C. This is the system most widely used in tenement construction. The old plasterwork is stripped from the walls to a height of

1200mm. Holes are drilled into the wall 100mm apart along the line of the new D.P.C. level. The waterproofing chemical is then fed into the holes to impregnate the wall structure and create the new D.P.C.

There are **water-based** and **solvent-based** chemicals. The water-based chemicals can be fed into the wall using a gravity method. This can take some time but is easy to monitor. The solvent

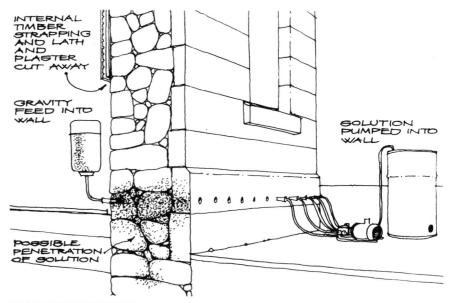

18.2 DAMP PROOF COURSE INJECTION

based chemicals can be fed into the wall under pressure, a quicker process. The chemicals used can be quite toxic so it is best to leave the room unoccupied and ventilated once the installation is complete.

The wall is then replastered to a specification set down by the specialist. **Cement render** is commonly used but there are other specialist plasters on the market. On **no** account should ordinary **Gypsum plaster** be used as this can become hygroscopic resulting in moisture being retained in the plaster. The wall may not dry out completely for nearly a year afterwards. Indeed the cement render is applied largely to allow decoration to proceed as moisture in the wall cannot penetrate easily through the render.

Solum

In severe cases of rising damp it may be necessary to remove the old flooring timbers. The space underneath is called the **solum** and should provide a ventilated space to prevent timber rot. This space should be cleaned out and sterilised. It should be covered in

100mm of clean gravel. Sometimes it is treated with hot bitumen and the bitumen is taken up the sides of the wall (this can prevent this section of the wall from drying out adequately). In all cases the solum area should be well ventilated, with grilles through the front and rear walls. The **dwarf walls** (also known as **honeycomb walls**) supporting the floor joists should also be well ventilated.

What to Look Out For

Water-based silicone treatment is evidently cheaper than solvent - based silicones. The water-based silicone when injected into the wall will tend to dilute with the existing water in the wall. If the rising damp is serious, there is a tendency for the chemicals to spread throughout the wall, thus reducing its effectiveness.

The other main areas to look out for are:

Holes may not be drilled deep enough into the wall to dispense the silicone. In a 600mm thick wall the injection has to be done in stages and at different depths.

Water based silicones have to be diluted. If not controlled carefully over-dilution can occur.

Solvent chemicals have to be injected to the correct pressure and volume. Once again site control is essential at this stage.

The D.P.C. must be correctly placed. In some cases a vertical D.P.C. is required but omitted from the process.

Penetrating Dampness

This is the most common cause of dampness: water penetration through the roof, wall, doors and windows. Problems with these elements are dealt with under their own sections but some of the more common problem areas are listed below:

Stone walls which are constantly getting wet through defective gutters and downpipes, allowing water to penetrate the wall and cause rot in the timberwork inside.

Cracked or decayed stones, or poor pointing can all allow moisture to penetrate the wall.

Poor flashings and slipped slates or tiles may be difficult to spot without close inspection.

Inadequate window replacements often cut out the timbersub-cill allowing water penetration under the cill.

Dampness From Plumbing Defects

In most Victorian tenements, the original pipework supplying water to the loft storage tanks and fittings would be lead. In addition, lead waste pipes may be used to connect fittings to cast iron waste pipes. Over time, this leadwork will deteriorate and may cause a very small leak, hardly detectable but enough to start rot in the flooring timbers.

Timber Decay

If wood becomes damp it will support a variety of fungal infections, but no fungus will propagate in the first instance if the moisture content in the timber is less than 20%. The temperature range in which the fungus grows intends to restrict it to temperate climates and as a result it is more commonly found in Britain than anywhere else.

Some of these infections are superficial and will not affect the strength of the timber (i.e. fresh green timbers can get a blue stain on it known as **sapstain**).
If wood remains continuously wet but is still exposed to the air, a soft rot fungi is likely to develop, this is called **wet rot**. If the wood is slightly drier and at the correct temperature and humidity, a more serious form of rot can develop, known as **dry rot**. Woodworm outbreaks will also lead to serious damage to timbers. This section describes these three problems, how they can be surveyed and how treated.

Dry Rot
(Serpula Lacrymans)

Once dry rot fungus has become established (preferring a moisture content between 20% and 35% and a temperature between 50°C and 260°C), it is able to generate moisture through the digestion of wood and by so doing can spread to timber which has a moisture content of only 14%. It is also able to maintain the correct atmosphere in poorly ventilated conditions. This is the reason it is found in unventilated spaces - behind wall panelling, under floorboards, skirtings etc.

18.3 EXTENSIVE DRY ROT OUTBREAK

The fungus **mycelium** develops on the surface of the infested timber forming a white to yellow 'cushion'. Feeder strands or **tendrils** then develop within the mycelium and supply water to the growing area. These strands can travel great distances and will pass through brickwork to reach new timber. **Fruiting bodies** called **sporophores** develop and discharge millions of spores for a two week period. These spores then infect the surrounding woodwork and growth is encouraged by the feeder strands.
The fruiting bodies may develop into a rusty red colour.
The decayed wood crumbles easily between the fingers and the wood may appear broken into small cubes along and across the grain.

Wet Rot
(Cellar rot or Coniophora Puteana)

This requires consistently damp conditions with a timber moisture content between 50% and 60%. It is less affected by temperature than dry rot. It is usually found in semi-external conditions such as basements, window cills, rafter

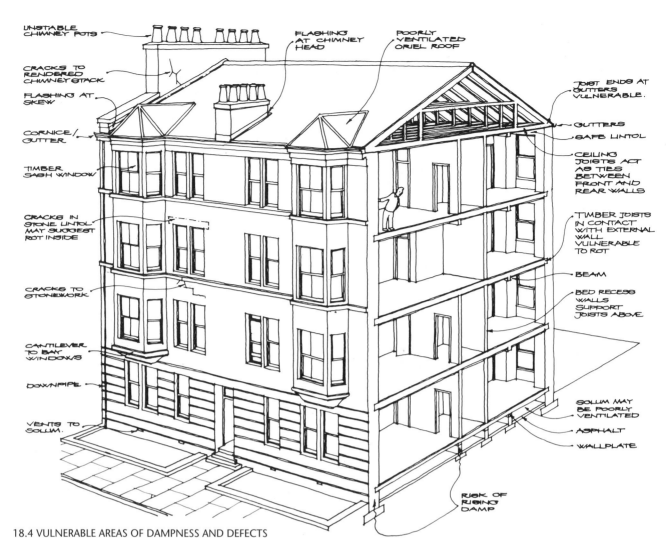

Labels on the diagram (clockwise from top left):

UNSTABLE CHIMNEY POTS

CRACKS TO RENDERED CHIMNEY STACK

FLASHING AT SKEW

CORNICE / GUTTER

TIMBER SASH WINDOW

CRACKS IN STONE LINTOL MAY SUGGEST ROT INSIDE

CRACKS TO STONEWORK

CANTILEVER TO BAY WINDOWS

DOWNPIPE

VENTS TO SOLUM.

FLASHING AT CHIMNEY HEAD

POORLY VENTILATED ORIEL ROOF

JOIST ENDS AT GUTTERS VULNERABLE.

GUTTERS

SAFE LINTOL

CEILING JOISTS ACT AS TIES BETWEEN FRONT AND REAR WALLS

TIMBER JOISTS IN CONTACT WITH EXTERNAL WALL VULNERABLE TO ROT

BEAM

BED RECESS WALLS SUPPORT JOISTS ABOVE

SOLUM MAY BE POORLY VENTILATED

ASPHALT

WALLPLATE

RISK OF RISING DAMP

18.4 VULNERABLE AREAS OF DAMPNESS AND DEFECTS

ends, or parapet gutters. The fruiting body it produces is rarely found inside a building although it is quite common out of doors. The infected wood becomes very dark brown, and cracks appear along the grain of the wood.

Both wet and dry rots are capable of producing almost identical failures in wood. It is important not to over-rely on timber grain cracks in distinguishing between the two types of outbreaks.

Specialist Surveys

Most 'specialist' wood treatment firms will offer a service for the inspection of rot. In many cases the advice given is not of very high quality as employees may be unqualified, the firm relying on obtaining the eradication work, rather than depending on the quality of the survey. However a full technical survey for rot can cause a lot of disturbance. Floorboards as well as carpets may need to be lifted, walls may need opening up

What to Look Out For

All rots are associated with a source of moisture. Any obvious defect which is likely to lead to water penetration inside should be thoroughly inspected for rot. The following areas are vulnerable:

Wet rot is commonly found at the seating of **rafters** and **ceiling joists** as well as around chimneys. If the ties between rafter and joist are affected then the roof may sag downwards, pushing out the stonework at the eaves.

Cracked or defective stonework combined with poor maintenance can allow water to penetrate the stonework and allow rot to start on the joist ends. The most vulnerable

areas tend to be where the plumbing stacks are, at kitchens and bathrooms.

If external lintols are cracked it may suggest that the internal timber **safe lintol** is rotten and the weight of the wall is being transferred to the outer lintol.

If floor joist ends are rotten, the tying effect between wall and floor may lessen to the extent that the outer wall starts to bulge outwards.

Rising dampness is often associated with wet rot particularly at the wallplates and joist ends touching the wall. A change in temperature and ventilation of the solum area can lead to dry rot outbreaks. Indeed it is often

thought that increased heating and insulation could in some cases exacerbate the dry rot problem.

Condensation in flat roofs can provide enough moisture to cause both dry and wet rot. Many small roofs over oriels are badly ventilated and prone to this.

Timber windows will become infected with wet rot if they are not regularly painted. Some window replacements do not involve the replacement of the old window frame. The old timber cill is removed and a ply facing applied. Water can then find its way under this cill and cause rot outbreaks below the window.

to inspect hidden timbers and lining boards and decoration can be damaged in the process.

As a full opening-up survey is rarely acceptable, only the most vulnerable areas of the building may be inspected *(see fig 18.4)*. The surveyor can also make use of a borescope which allows the inspection of hidden cavities without the need to open up floorboards.

A novel way of locating dry rot is to use *Rothounds*. These are dogs which are trained to sniff out any active dry rot in a building. They can also go into spaces which are too small for a human. Suspect areas identified by the dogs can then be inspected using borescopes, so the disruption is kept to a minimum, yet the accuracy of finding the rot is very high *(see Hutton and Rostron Environmental Investigations Ltd who use the rothounds).*

Remedial Work

Since all rot is caused by some form of dampness the first priority is to repair the defect which permitted the dampness. Withdrawal of the source of moisture will kill wet rot and affected timbers need only be cut out and replaced. However if dry rot is encountered the remedial work is more extensive. All infected timbers must be removed off site and even apparently good timber in the near vicinity will have to be replaced because of the possibility of spore infection. If joists are being replaced then the sound deafening will also need to be renewed. The surrounding walls will have to be sterilised and the new timbers treated with preservatives.

Because dry rot is not stopped by brick walls, neighbouring flats will need to be inspected. If there is dry rot next door as well and your neighbour ignores it, your flat will be at risk from further infections. If this happens it may be prudent to treat the mutual wall with a **toxic barrier** like **zinc oxychloride plaster.**

The preservative treatments to walls and timbers should be carried out be a specialist sub-contractor who is preferably a member of the *British Wood Preserving Association*. In order to reduce the harmful effects of the chemicals, resinous sawdust should be left on the treated surfaces for two days in order to absorb the residue chemicals. The sawdust should then be vacuum

cleaned and the exposed surfaces washed down with **Gloquat** to wash off any residual preservative. Try not to re-occupy the house for at least a week, although the time will depend on the type of chemicals used and the extent of the treatment.

The Green Approach

Because Building Societies often insist that any remedial work done in connection with dry rot has a 30 year guarantee, there is a tendency to specify fairly drastic remedial measures when a simpler (but not guaranteed) repair would suffice. In fact most Building Societies will accept work which is specified by a professional, even though there may not be a 'guarantee'.

There is a growing body of professionals who believe that it is the environmental conditions which require control, and that chemicals and drastic structural repairs cause more problems than the

symptom they are trying to cure.

It is known that treated timbers and sterilised walls can still become rotten, if the right moisture and humidity conditions are present. If the old rot is removed and the conditions which created the rot altered, then it should not recur. Treated timbers are not therefor required (indeed timbers which are treated with preservatives can be hazardous if they are burnt, as toxic fumes are given off).

If timbers are hidden from sight, then they can be fitted with electrical capacitance monitors which will keep a track of any changes in the moisture condition of the timbers (see Hutton and Rostron's 'Curator' system).

Toxic Chemicals

Wood preservatives are quite toxic and can remain active for 20-30 years. People can become ill after handling treated timber, or breathing spray mist, or the vapour left over after treatment. Several of the most commonly used chemicals cause cancer, allergy, nerve damage and damage the immune system. We list below some of the main chemicals which have been used and their side-effects. *For further information on this subject see: Toxic Treatments; Wood preservative hazards at work and in the home. 1988 The London Hazards Centre Trust.*

Arsenic

Only used in pretreatment, usually with copper and chrome but also in timber. Treated timber is most dangerous in the first two weeks after treatment. (Poisonous, causes skin damage and other cancers: can damage peripheral nerves).

Creosote

Common DIY product which is banned in the USA for all but professional use. (Highly poisonous, causes skin and eye irritation, nausea, headaches, bronchitus from spray,

and cancer of skin and lungs).

Dieldrin

Has been used in the remedial treatment industry but has now been banned under EEC directives. (Highly poisonous, contaminates through skin to affect nervous system; causes cancer).

Lindane

Used largely as an insecticide, but has been used for pretreatment work. Banned in many countries. (Highly poisonous, poisons through the skin, affects brain and nervous system, causes epilepsy, causes cancer in animals).

Pentachlorophenol (PCP)

Used for pretreatment and remedial work. Found in some DIY wood preservatives. Banned in many countries. (Highly posonous, poisons through skin causing nerve damage, paralysis, skin rash and even death)

Tributyl tin oxide (TBTO)

Used for pretreatment and remedial work. Banned as a boat anti-fouling paint because of its affect on marine life. (Highly posonous, absorbed through skin. Burns the eyes and skin, likely cancer risk).

What Your Factor's Contract Says in Plain English

This is one contract drawn up by the Property Managers Association Scotland Ltd.

1. In this contract and conditions, the term owner refers to anyone who owns a flat whether they occupy it themselves, let the flat out or leave it empty. A flat is any part of the property which can be occupied as a self contained dwelling.

2. The factor appointed by the owners will manage the whole property. This does not affect your rights as an individual owner and does not give the factor the right to do things which individual owners are not allowed to do. (a)

3. This contract gives the factor the authority to order common repairs ·where the estimated cost is less than £...... This sum may be changed if both factor and owners agree. If the repairs are estimated to cost more than £...... then:-

The factor will send one or more estimates to the owners.

A majority of owners will need to agree to the repair.

The owners must pay the factor the whole cost of the repair in advance before he will go ahead and order the repair.

If the factor believes that repairs are necessary on the grounds of health and safety, he will order these repairs, regardless of the conditions above, and recover the money from the owners afterwards. (b)

4. Every owner will deposit £...... with the factor as an advance payment. This sum can be changed if both factor and owners agree. No interest will be paid on the deposit. The deposit must be paid immediately an owner buys a flat and will be returned when he sells. (c)

5. Each owner's share will be that laid down in the Title Deeds. Each owner will pay their share promptly. If an owner does not pay their share, the factor will be entitled to sue them in the name of the other owners and obtain a court order requiring them to pay. If the owner does not pay within 21 days of the Court Order, the factor will be entitled to recover his share, plus expenses, from the other owners. These owners will then be entitled to take the non-paying owner to court. (d)

6. When an owner sells their property, they must tell the factor the new owner's name and the date the sale went through. The existing owner must try to persuade the new owner to adopt this contract.

7. The factor's appointment may be terminated by either side on three months notice.

8. Decisions about the common property and related matters will be taken in the manner set out in the Title Deeds. If there is no Title Deed, or if it is generally ignored then decisions will be taken by a majority of owners. Any decision taken will be binding on all owners.

9. Unless the Title Deeds say otherwise, each owner will have **one vote per flat owned**.

10. The Factor will ensure that the property has a **common insurance policy**. Owners will be expected to pay their share of insurance premiums immediately they become due. The factor will have no other responsibilities in connection with the insurance policy. The terms of the insurance policy and the amount of cover will be decided in accordance with paragraph 8. (e)

11. These conditions will apply to any owner who signs the contract, even if other owners in the tenement do not sign.

12. Attached to the factor's contract will be a **Schedule of Duties** which lists the jobs which the factor is responsible for. However, the factor will not be held liable if he fails to order repairs on his own initiative following a visit to the property. (f)

POINTS TO NOTE

(a). This says that you can only have one factor in each tenement.

(b). Most factors have set the sum at £200.

(c). This is the notorious float which many people have refused to pay. Most factors have set the float at £30. If the interest rate is 10%, you will 'lose' £3.00 a year in interest.

(d). This may seem very unfair but it is generally held that as all owners have benefited from the work, it is right for them to pay for it, rather than the factor or tradesmen.

(e). If you have a claim against the common insurance, you will need to sort out everything by yourself.

(f). There may seem little point in the factor visiting the property if you can't rely on him to order repairs where appropriate. On the other hand, not even a trained surveyor will guarantee that he has found all the problems in a property following a full structural survey.

Appendix 2

Schedule of Factorial Duties:

A factor's contract should be accompanied by a schedule of duties outlining what a factor will do. The following schedule of duties accompanies the factor's contract drawn up by the **Property Managers Association Scotland Ltd.**

1. "He will make **periodic** visits and take appropriate action to deal with any matters of a common or mutual nature which are discovered." (a)

2. "Unless you tell him which firm to use, he will order repairs from a firm he believes will give good service and a reasonable price. He may ask the tradesman what kind of repair is needed and what materials should be used." (b)

3. "Where more than one trade is required he will arrange for the different firms to co-ordinate their work. Where he considers it to be in the best interests of the owners he will obtain several estimates for the same job, advise the owners and obtain their instruction before proceeding." (c)

4. "Where the repair is mutual to an adjoining building he will negotiate with the adjoining owners or factor and endeavour to ensure that the adjoining owners pay their share of the cost. He will not instruct such work without such agreement unless he is so authorised by his owners." (d)

5. "He will investigate any complaints of **unsatisfactory work**. Owners will assist here by reporting any such complaints to the factor as soon as possible. Where considered necessary and if so instructed by the owners the factor will arrange for a professional report on the completed repair. Fees for this report will be chargeable to the owners. He will check the Tradesmen's Accounts when rendered, including the charge of VAT, will calculate the share of the cost due by each owner in the building and unless otherwise agreed will issue half yearly accounts to each owner." (e)

6. "He will also ensure that accounts for **ground burdens**, insurances and all other outgoings are checked and paid when due. He will calculate the shares due by each owner and include same in the half yearly accounts."

7. "On appointment the factor will agree the **management charge** with the owners and such charge will be revised as necessary from time to time thereafter. Such revisions will not normally be made more frequently than once a year. The management fee as agreed from time to time will cover routine management duties but it is understood that if because of the complexity of a particular repair or because of any other reason the factor is involved in extra work an additional fee may be chargeable." (f)

8. "When a change of ownership takes place the factor will on request make the necessary apportionment of ground burdens, insurances, repairs and other outgoings between the seller and the purchaser. Any charge for this additional work will be payable by the seller. **If requested** he will arrange to supply one photostat copy of the Tradesmen's Accounts to one owner per building each half year." (g)

9. "He will guide and assist owners in submitting **applications for grants** towards the cost of common repairs or improvements."

10. "In the event of any court action being raised on behalf of the owners by or against any third party instructions will first be taken from the owners as they will be liable for all legal costs not recovered." (h)

POINTS TO NOTE

(a) 'Periodic' means something which happens regularly. If it's a regular 10 years, it still fits the clause but it's not much use to you.

(b) This section says that a factor will use firms of their own choosing and is not obliged to get estimates for every job. In many cases, the factor will reply on the tradesman for information about what repairs are needed. It doesn't give the factor a lot of control and there is nothing which says the factor will check repairs.

(c) It is hard to know what the factor considers your best interests to be. You can instruct the factor precisely when you want several estimates obtained.

(d) This refers to a repair which has to be paid for by two sets of tenement owners. If the other owners are being uncooperative your own factor will get the work done but he will probably want you to agree to pay the whole cost.

(e) A factor will only do something about a poor job if you tell them. You must check repairs yourselves. You won't know for sure that the job is complete until you get the bill some months later from your factor. By this time it may be too late to check properly. If a factor isn't checking the work, they can't very well check that you have been properly charged.

(f) This suggests that the factor should consult with you before putting up the management fee or charging extra. It is difficult to know when the factor will decide that extra fees should be charged. You could ask them to tell you before the fees are incurred.

(g) Again, this is something not included in the management fee. Note the 'if requested'. You may have to go to the factor's office to get the photocopy. This is a service you will probably be charged for.

(h) The schedule does not say what kind of help you will get from your factor when it comes to going to court. The chances are you will be charged for any help you get.

Appendix 3

Stair Association Model Rules

Use these rules as a guide. Rules marked in **bold type** will need to be checked against your title deeds.

Strike out the options you don't want after discussion with your neighbours.

1. All owners will be members. There will be one vote per flat.

2. The Association will meet at least twice a year, more often if necessary. Owners will be given 7 days notice of a meeting. The owners of any 2 flats may call a meeting.

3. Decisions will be taken by a majority vote taken at the meeting.

4. A record of decisions will be kept and absent owners will be notified of decisions after the meeting.

5. Each person will carry out the tasks given to them at the meeting and will have the authority to act in accordance with decisions taken at the meeting. Tasks will be spread amongst owners so that each owner makes a fair contribution to the running of the Association. Absentee owners will be expected to contribute £..... a year to the running costs of the Association in lieu of doing work.

OR

All owners will act as factor arranging repairs and collecting money on a rota basis, each period of duty lasting for one repair or............... months.

OR

The Association will elect the following office bearers annually:-

> *Chairperson or factor*
>
> *Treasurer*
>
> *Backcourt organiser*
>
> *Correspondence secretary*
>
> *Staircleaning organiser*

6. The Association will keep a bank account at

_____

_____

_____

7. Cheques and other withdrawals must be signed by at least two of the following people

_____

_____

_____

8. All owners will make a regular contribution of £ ..... a month to Association funds. If the Association fund stands at over £....... the Association can decide to refund money to members.

OR

The Association Account will be maintained at a minimum of £.....

When the account falls below this amount, all owners will pay their share of the difference.

9. The Treasurer will keep a petty cash fund. The petty cash will never be more than £..... Owners will be refunded cost of phone calls, bus fares, stamps, etc. where these costs have been incurred in arranging repairs or running the Association. The Treasurer will keep a record of money paid out of the petty cash fund.

10. Association funds will only be used to pay for common feu duty, common insurance, backcourt maintenance, common repairs and the running costs of the Association. (Add or delete items to suit your own stair).

11. Common repairs are as follows:-

_____

_____

_____

_____

_____

_____

_____

_____

12. Common repairs will be paid for equally by all owners, or common repairs will be paid pro rata according to rateable value of flats.

13. All common repairs will be organised by the Association. Unless there is an emergency, the Association will not be bound to pay for repairs not arranged by them. In an emergency the following firms may be called:-

_____

_____

_____

_____

If the emergency is a burst pipe or tank, water will first be turned off at the street toby and a plumber called only if the emergency continues. This is to prevent an unnecessary out-of-hours call-out charge.

14. Whenever possible the Association will get three estimates for repairs.

15. All repairs over £..... in cost must be agreed at a meeting before an estimate can be accepted.

16. All owners must pay in advance for repairs over £.... Money paid in advance will go into the Association's account.

17. Any owner putting a flat up for sale will draw the rules of the Association to the attention of prospective purchasers, and their surveyors and solicitors.

18. When an owner sells, they will be reimbursed their share of accumulated funds, minus the cost of any repairs ordered before their date of removal.

19. All interest accumulated on the account will be spent on common items and not returned to individual owners.

20. All owners will be given a copy of these rules.

Signed by all owners.

Appendix 4

Stair Association Minutes of Agreement

1st Minute of Agreement

Meeting of Stair Association held on
..../..../1992.

At a meeting attended by

_____

_____

all owners of flats at

_____

the following was agreed:-

1. Our tenement requires the following repairs:

_____

2. The following owners/all owners attending agreed:

(i) That they intended to pay their share of repair costs, providing the final cost does not exceed £....

(ii) That

_____

and

_____

will act as close representatives obtaining estimates, applying for grants and hiring architects/surveyors.

3. This minute does not give our representatives the right to make changes to the proposed repairs scheme or to accept any estimates, tenders, etc, without first getting the agreement of all/the majority of the other owners.

4. As the meeting was quorate with the required number of owners attending and as the number of owners agreeing to the above decisions forms a majority, the decisions taken will be binding on all owners of flats in this property.

5. Signed by

_____

2nd Minute of Agreement of........Stair Association

* *delete as appropriate to your title deeds*

At a meeting attended by

_____

_____

_____

all owners of flats at

_____

it was agreed that:-

1. The following common repairs would be carried out to our tenement:

_____

_____

2. All owners understand that not all common repairs are eligible for grant and that some work to some flats is entirely the responsibility of the individual. Some of the individual repairs may be grant aided.

3. The share of common repairs and the individual repair costs for each flat are set out on the following pages. **All signing owners agree to pay their share of these costs.** As these owners form a majority according to the provisions of the Deed of Conditions, the decision to carry out the above common repairs will be binding on all owners.

4. The work will be carried out by

_____

and supervised by

_____

5. Any changes to this agreement such as:-

major changes in running the contract

major changes in the repairs themselves

any changes which involve additional costs of more than ..% of the total MUST be agreed before hand:-

* *by all owners in writing, or*

* *majority of owners in writing, or*

* *at a meeting of owners*

There will be one vote per flat.

At the owners' meeting, a decision must be agreed:-

* *by a majority of all owners, or*

* *by a majority of those owners that attend the meeting*

All owners agree to abide by any decision made in the way set out in this agreement and agree to pay their share of any increased costs resulting from this decision whether or not each owner agrees with the decision.

6. Any owner may call a meeting to discuss the progress of the works on three days' notice.

7. Our representative(s) will be

_____

These representatives may be voted out

* *by a majority of owners in writing*

* *by a majority of owners at a meeting*

The close representatives are authorised to carry out the following on our behalf:-

accept tenders

instruct additional works in accordance with this agreement

appoint professional or other advisers

take day-to-day decisions over the management of the contract

operate a bank account on our behalf

apply for and receive grants on our behalf

make complaints to the contractor

negotiate damages/remedial works with the contractor

carry out any other tasks incidental to the completion of the works.

All owners agree not to ask any workmen, professional advisers or council officers to make any changes to the scheme without first consulting the close representative.

8. All payments made by owners will be paid into a separate account and all withdrawals on this account must be signed

*by

_____

*any 2 of the following owners

_____

_____

All owners agree to pay £................. in advance of work starting and within .... days of being given notice by the close representative.

All owners agree to pay any reasonable sum required by a close representative on .... days notice,

OR

All owners agree to pay on ...... days notice of a decision to pay being made at a meeting.

All owners understand that it will be necessary to pay the contractor in full without first receiving grant money.

9. All owners agree to give access to my/our flat on..... days notice or to leave instructions with the close representative on how and when access will be given.

All owners agree to move furniture and carpets in advance of workmen calling. All owners will inform the close representative if they cannot do this.

10. If any owner decides to sell their flat, they will pay the appropriate share of the cost of the works themselves unless the new owners agree to sign this contract and take responsibility for paying the repair costs.

All owners agree to inform the close representative on putting the flat up for sale and give the new owner's name and address.

All owners agree to show this contract to all prospective buyers and their legal representatives.

11. This agreement will not come into force unless it is signed by

> * *all owners, or*
> * *the majority of owners*

This agreement will automatically come to an end when repair works (incl. defects) are complete and paid for.

12. All owners agree to co-operate fully with other owners in getting work properly and quickly carried out.

Signed

_____

Date

_____

INSTRUCTIONS

You must get the right person or people to sign. If there is only one individual owner you will need one name and one signature at the end. If there is more than one owner of a flat you should put down all the names and get all the owners to sign. If you are dealing with several owners or a limited company you can ask one person to sign 'on behalf of' the other owners - but that person must be authorised to act on behalf of the other owners.

In paragraph 1 and 3 give as many details as you can or refer to the specification or estimates that you have received e.g. *'as set out in Bills of Quantities drawn up by Lines & Co. Architects dated 20 May 1993'.*

It is very likely that you will discover additional works are needed. You should include a figure in paragraph 4 like 5% or 10%. If you make it too low you will have to have a meeting every time you make a minor change.

You should discuss how you will agree to changes with other owners before filling in this agreement. Be guided by what your title deeds say.

You may not need all of section 8 if you already operate a **stair account.**

Appendix 5

The Standard Building Contract

There are two main types of contract, the 'JCT' and the 'Scottish Minor Works Contract'. The minor works contract does not normally use Bills of Quantities but otherwise the two contracts are very similar. This is what the contracts say:

1. Costs

The price given by the contractor is the **fixed price** of the contract but changes made during the contract can affect the final price of the works. These changes must be agreed by both the architect and contractor. The architect should refer big changes to you. Costs can change because of the following:-

Provisional Sums - These are estimates included in the contract for work which cannot be easily measured beforehand. The final cost for any item may be more or less than the provisional sum.

Dayworks - Provisional sums and extra items are often measured as dayworks. These rates are for the time spent on the job plus the cost of materials. Your architect will try to avoid dayworks as the rates tend to be very high. The contractor can also work as slowly as they want and still get paid. In other parts of the contract the contractor gives a price per job - the contractor is then paid the same amount for the job whether they take one hour or ten.

Prime Costs - These are the costs allowed for using a sub-contractor to do specialist work or to supply special goods. Sometimes the actual cost may be different.

Variations/Architect's Instructions - These are instructions given by the architect which alter what was put in the contract. If additional work is instructed, you will have to pay for it. These instructions are also used to confirm provisional sums.

Loss & Profits - If there are delays on the job you may have to pay the contractor's extra expenses and any profits they may have lost by not being able to do other jobs.

Contingency Sum - Usually about 5% of the contract value. This sum is designed to cover the cost of unforeseen

work.

Savings - If the price of the lowest tender is too high, the architect may propose and negotiate savings. This can be done by either cutting down the contingency sums (and hoping that too much additional work won't be needed) or by reducing the specification *(for example, repointing a chimneyhead rather than completely rebuilding it)* or by cutting out some items of work altogether.

Claims - Both you and the contractor may make claims which are added to the contract sum. You may claim for **liquidated damages** if the contractor does not finish in time. Liquidated damages can only cover actual costs arising from the delay. It may include a small amount for inconvenience but it generally only covers the additional cost to the architect of inspecting the contractor for a longer period. Most claims arise through **extensions of time** but another major cause for claims is additional work caused by unforeseen problems or you changing your mind. You may also get a claim for time lost trying to get access. You may go to arbitration about claims. The contractor may claim for delays in obtaining information from your architect.

Remeasurement - If more or less work has actually been done than was stated in the bill, the quantity surveyor may remeasure the work. The contractor will be paid pro-rata.

2. Payments

The contract normally states that the contractor must be paid within two weeks of a certificate being issued by the architect. If you do not pay on time, the contractor may cease work on the contract or charge you interest.

3. Quality of Work

The architect is responsible for inspecting the work, usually on a weekly basis. If a closer check on work is required either you or the architect can employ a clerk of works. This will usually cost an additional 1% or 1.5% of the contract price. If you employ a clerk of works you can expect to get a weekly report.

It is up to the contractor to organise and programme the work, the architect or

clerk of works cannot tell the contractor **how** to organise it. They can only make sure the work complies with the contract requirements. They can ask the contractor to open up work which may be defective but, if the work turns out to be alright, the contractor has to be paid for making good what has been opened up. The contractor is responsible under common law and contract law, for poor workmanship or low quality materials.

The defects liability period allows problems to come to light. The contractor is still owed **retention** money and you do not pay this money until the defects are put right. If the contractor won't come out to deal with the defects the architect can call in another contractor and take the cost from the retention money, but this is not always enough.

The contractor cannot argue that the defect should have been spotted before and they must put it right. Even after the end of the defects liability period, the contractor can still be held responsible for latent defects, such as plaster shrinkage, but not repairs caused by storms or TV aerial men knocking tiles loose. As you will have finished paying the contractor you may not have much success in getting latent defects put right.

4. Stages of Work

The following comments refer to the JCT Contract (1983) only. The Scottish Minor Works Contract is less precise in its terms but tends to follow the same pattern:

(i) The first date given in the contract is for site possession or site start. This is normally set for some 2-6 weeks after the contract has been signed. It allows the contractor to finish off other jobs and get material, scaffolding and workmen. If the contractor does not turn up, the architect (with the client's authority) can get the next lowest tenderer to do the work.

(ii) After this the contract is "on site". The contract period is usually set out in the Bills of Quantities and the contractor is bound by this time period. If the contract is to last longer than one month, you will probably have to make stage payments. A payment is normally made every month. The architect or quantity surveyor will make an interim valuation of the work done to date and the architect will issue an interim certificate. The certificate says how much must be paid. An agreed percentage (usually 2.5% - 5%) of the cost is set aside to be paid later when defects have been rectified. This is referred to as retention money.

(iii) Practical Completion is when all the work has been done except minor items, which can still be seen to later, and defects. The certificate of practical completion allows the contractor to be paid for almost all the contract price except for half the previous retention which is held until the end of the defects liability period. From the date of practical completion, the contractor is no longer liable for any damage to the work unless it is due to a latent defect which does not appear for some time.

(iv) The Defects Liability Period starts on practical completion and usually continues for 6 months or one year. During this time, the contractor must deal with any urgent defects. At the end of the defects period, the architect will inspect the work and send a list of defects to the contractor. These defects cover items such as shrinkage cracks, leaks at plumbing joints, spalling render, stains in paintwork etc. When the defects have been rectified the architect will issue a certificate of **making good defects**. The retention money will now be paid to the contractor.

(v) After the defects have been completed, the quantity surveyor will draw up the final account, and the architect will issue the **Final Certificate**. This is the end of the contract but either you or the contractor can query the Final Certificate within 14 days.

Extension of Time

At any stage, the contractor may ask for an extension of time on the following grounds:-

Exceptionally bad weather, vandalism or other damage, war or strikes.

Architect's instructions or variations.

Delays by nominated sub-contractors.Delays caused by re-inspections.

Delays caused by the architect not sending information.

The contractor can claim extra payment for all of these delays except bad weather which cannot be the source of claims for

either side, that is you cannot claim liquidated damages for weather delays.

5. Your Liabilities and the Contractor's Liabilities.

Both you and the contractor are responsible for your own negligence. *For instance, you are negligent if you climb the scaffolding without the contractor's permission.* You may also be asked to pay for any damage that you cause to the scaffolding. The contractor is negligent if they fail to remove ladders from the scaffolding over the weekend and a young child climbs up and falls. The contractor is also responsible for sloppy working or accidents caused by the workmen. The architect is responsible for the design and specifications but the contractor has an obligation to inform the architect about obvious mistakes. If there is vandalism or theft which is not the employer's (your) fault before practical completion then the contractor should put things right, at his own cost.

6. Insurance

In order to protect yourselves, both you and the contractor are obliged to take out insurance. The contractor insures against their own negligence and liability. You must arrange insurance against fire, storm, burst pipes or earthquakes. This insurance must cover any damage caused by these things including fires which are due to the contractor's negligence. Your architect will usually allow for such insurance within the contract. Sometimes the architect may recommend that you take out additional insurance, but this will depend on the type and extent of work.

The contractor is responsible for leaving the work reasonably secure. They should not leave windows open and should not make access to the scaffolding easy. However, the contractor is not responsible if you do have a break-in despite these precautions, and you will have to claim on your own contents insurance.

You should inform your insurance company that work is taking place to ensure you will be covered.

The contractor is also responsible for leaving the work reasonably well protected from the weather. If the roof work is not completed, it should be left covered by a tarpaulin at night. If there is a storm which blows off the tarpaulin you will have to claim on your building

insurance for any storm damage to your flat, unless a common policy has been taken out specifically through the contract.

7. Sub-contractors.

A sub-contractor is a separate firm which works with the main contractor on a special part of the work. If you or your architect ask for a particular sub-contractor, perhaps to install a communal aerial, or carry out structural work, this is known as a **nominated sub-contract.** You are responsible for many aspects of a subcontractor's work but the main contractor should check the quality of the sub-contractor's work. If the contractor appoints a sub-contractor, perhaps because there are no electricians in the **main firm, this is known as a domestic sub-contract.** The main contractor is responsible for all aspects of the domestic sub-contractor's work.

8. Problems

Your architect will try to resolve most problems with the contractor. If the problems cannot be resolved you can **determine** the contract. This means the contract will be terminated. The architect will be able to advise you of the consequences of such action.

Alternatively you can go to **arbitration.** This more suitable for a contract which is finished in the sense that the work has been done. One party may pursue arbitration if the other does not agree to. The contract should state who will nominate the arbiter.

You can determine the **JCT Contract** if:-

The contractor unreasonably stops work,

The contractor does not work regularly or diligently (the architect must give the contractor 7 days' notice of their intention to determine the contract).

The contractor refuses to rectify defects when instructed to do so by the architect.

The contractor goes bankrupt.

The contractor can determine a contract if:-

You do not pay them (they should give 7 days' notice of this).

There has been damage which is not the contractor's fault.

Work is delayed because the architect

has not sent out instructions in time.

The architect insists on carrying out unnecessary reinspections.

You interfere with the work or do not give the contractor access to the premises.

You suspend the work for at least one month.

You go bankrupt.

The contractor must give you 7 days' notice of the complaint before determining the contract.

Under the **Scottish Minor Works Contract**, you can determine the contract if:-

The contractor does not work regularly or diligently.

The contractor does not finish the work.

The contractor goes bankrupt.

If you determine the contract you may claim for additional expenses in getting a new contractor on site **but** you will have to pay for any work the first contractor has done. If the contractor determines the contract, they will charge you for the work they have done and can also claim damages for any losses they have suffered.

WARNING
Determining a contract is the last thing you should do. It is legal but not advisable. There will be delays of six months to one year before you can get another contractor on site. The new contractor invariably condemns the previous contractor's work (for which the new contractor is expected to assume responsibility). If the work is condemned, it will have to be redone and you will end up paying twice over.

Appendix 6

Agreements for Building Work

If you are intending to appoint a builder to carry out minor repair work, you may do so on the basis of accepting his quotation, which, together with your specification and request for the work, would become the contract between both parties. You may decide to use a very simple pro-forma contract such as the Home Improvement Contract Kit which is suitable for small repair contracts. If you are intending to carry out work in excess of £1000, you should consider using The Scottish Minor Works Contract (Jan 1992 edition), prepared by The Scottish Building Contract Committee. You should certainly consider using an architect or surveyor if the work is major, or involves planning or building warrant applications.

If you are intending to appoint a builder yourself then try to ensure the following items are included in your request for a quotation.

1. Contacts

State who will be responsible for liaising with the contractor. You should give at least two names.

You should state who will be responsible for providing access and if there are any restrictions on the access.

Ensure that the builder provides you with an emergency phone number where you can contact him if something goes wrong, particularly out of normal working hours.

2. Working Arrangements

Set down the frequency of any progress meetings you feel will be required with the builder.

Set down the times within which the builder can work on site. i.e. 8am to 6pm, not on Sundays.

Set down any particular conditions and ensure you ask the builder to remove all debris and rubbish at the end of every working day and leave the premises reasonably clean at all times.

Advise him of the need to satisfy all health and safety requirements to protect the safety of occupiers and other users of the building.

If it is a very small repair job, you cannot always expect the builder to provide his own toilet or washing facilities. However

if no-one in the close is willing to allow the use of their toilet you should state this so the builder knows he will need to make his own arrangements.

3. Programme

You should state when you want the work to start and how many working weeks you expect it to take. Ask the builder to confirm this as well as provide you with a statement on the order he intends to tackle the work.

4. Payments

Clarify when payment for the work will be required. For small repair jobs it is usual to make payment on completion of the work. You should explain who will be responsible for approving the work prior to payment. You should also define the amount of retention (if any) you intend to hold back. For small repairs a sum of 5-7% is normal, being held back for 6 months and released after a final inspection.

Assuming you have a stair association, you should clarify to the builder that the money will be held in a suspense account and state who will be responsible for making the payments on behalf of the owners. If you are dealing with larger sums of money, you should consider opening a trust account where the money is held in a joint account and can only be released after signatures from the builder and the close representative.

If you are asked to represent other owners and negotiate with the contractor, you should make it clear from the start that you are only acting as an agent for the proprietors. If you don't have a common account, the builder should have a signed agreement with each of the owners.

5. Variations

You can expect the builder to come back to you during the work asking for decisions on changes or variations he needs to make. Clarify with the builder who will make such decisions and confirm that any work which entails additional costs will not be approved until the builder has received a written instruction from your close representative. You should supply such a response within 48 hours or sooner so the builder is not delayed.

6. Insurances

The builder should carry public liability insurance for costs up to £1 million.

The builder should provide insurance for the works and insure against loss or damage by fire, lightning, explosion, storm, tempest, flood, bursting or overflowing water tanks and pipes.

The builder should indemnify the employer (all owners) against any expense, injury or damage whatsoever to property and contents, resulting from negligence or breach of statutory duty.

Ask to see copies of the builder's insurance certificates before you sign agreement to the work.

As the employer, you should instruct your insurance agent to adjust your common and individual insurance policies to take into account any risks associated with the work.

7. Sub-contractors

The smallest of repair jobs may require your builder to employ a sub-contractor. Clarify that the builder will be responsible for all actions including any work carried out by sub-contractors.

8. Approvals

The builder should obtain all necessary permissions and consents for the work to be carried out and completed. For example, siting of skips, approval of water board, electrical test certificates. However you should not expect your builder to apply for building warrant or planning permission.

9. Common Law Rights

State that none of the terms of your agreement with the builder removes common law protection relating to the quality of work or the reasonableness of payment for work, working arrangements or negligence by either party.

Further Information

Scottish Minor Works Contract available from RIAS bookshops. Home Improvement Contract Kit (1989) by K.Scott and D.Skelton, available directly from Architecture Design and Technology Press, 128 Long Acre, London WC2E 9AN.

Scottish Building Contract Committee, 27 Melville Street, Edinburgh EH3 7JF

Appendix 7

Contact Addresses

Age Concern

54a Fountainbridge,
Edinburgh EH3 9PT
Tel: 031-228 5467

Architects Registration Council of the United Kingdom

73 Hallam Street,
London W1N 6EE
Tel: 071-580 5861

Association of British Credit Unions Ltd.

Unit 307, Westminster Business Square,
339 Kennington Lane,
London SE11 5QY
Tel: 071-502 2626

Building Research Establishment

Kelvin Road, East Kilbride,
Glasgow G75 0RZ
Tel: 03552-33001

British Standards Institute

2 Park Street,
London W1A 2BS
Tel: 071-629 9000

British Board of Agrement

PO Box 195, Bucknalls Lane,
Garston, Watford,
Herts WD2 7NG
Tel: 0923-670844

British Insurance Association

19 St Vincent Street,
Glasgow G1 2DT

Citizens Advice Scotland

26 George Square,
Edinburgh EH8 9LG
Tel: 031-667 0156

Edinburgh New Town Conservation Committee

13a Dundas Street,
Edinburgh EH3 6QT

Energy Action Grant Agency

Bank Chambers,
9-17 Collingwood Street.
Newcastle Upon Tyne NE99 1NG
Tel: 091-2301830

Energy Action Scotland

21 West Nile Street,
Glasgow G1 2PJ
Tel: 041-226 3064

Glasgow West Conservation Trust

30 Cranworth Street,
Glasgow G12 8AG
Tel: 041-339 0092

Health and Safety Executive

Meadowbank House,
153 London Road,
Edinburgh EH8 7AU
Tel: 031-661 6171 and
314 St Vincent Street,
Glasgow G3 8XG
Tel: 041-204 2646

Historic Scotland

27 Perth Street,
Edinburgh EH3 5RB
Tel: 031-556 8400

Institute of Clerk of Works of Great Britain

41 The Mall,
London W5 3TJ
Tel: 081-579 2917

Institution of Structural Engineers

11 Upper Belgrave Street,
London SW1X 8BH
Tel:071-235 4535

Joint Contracts Tribunal

66 Portland Place,
London W1N 4AD
Tel: 071-580 5533

Lands Tribunal

1 Grosvenor Crescent,
Edinburgh EH12 5ER
Tel: 031-225 7996

National Federation of Savings and Co-operative Credit Unions

1st Floor, Jacobs Well,
Bradford BD1 5KW
Tel: 0274-75350

Office of Fair Trading

Chancery House,
53-64 Chancery Lane,
London WC2A 1SP
Tel: 071-242 2858.

Ombudsman: The Commissioner for Local Administration in Scotland

5 Shandwick Place,
Edinburgh EH2 4RG
Tel: 031-229 4472

The Parliamentary Ombudsman

Church House, Great Smith Street,
London SW1P 3BW
Tel: 071-212 7676

Planning Exchange

186 Bath Street,
Glasgow G2 4HG
Tel: 041-332 8541

Property Managers Association Scotland Ltd

2 Blythswood Square,
Glasgow G2 4AD
Tel: 041-248 4672

Royal Incorporation of Architects in Scotland

15 Rutland Square,
Edinburgh EH1 2BE
Tel: 031-229 7545

Royal Institute of British Architects

66 Portland Place,
London W1N 4AD
Tel 071-580 5533

Royal Institute of Chartered Surveyors in Scotland

9 Manor Place,
Edinburgh EH3 7DN
Tel: 031-225 7078

Scottish Federation of Housing Associations Ltd

40 Castle Street North,
Edinburgh EH2 3BN
Tel: 031-226 6777

Scottish Homes

91 Haymarket Terrace,
Edinburgh EH12 5HE
Tel: 031-313 0044

Shelter (Scotland)

53 St Vincent Crescent,
Glasgow G3 8NQ
Tel: 041-204 2154

Strathclyde Credit Union Development Agency

95 Morrison Street,
Glasgow G5
Tel: 041-429 8602

Scottish Development Department

43 Jeffrey Street,
Edinburgh EH1 1DN
Tel: 031-556 8400

Scottish Historic Buildings Trust

Saltcoats, Gullane,
East Lothian EH31 2AG
Tel: 0620-842757

Trading Standards Office

(check phone book for local offices).

1 St Enoch Square,
Glasgow G1 4BH
Tel: 041-204 0262

1 Parliament Square,
Edinburgh 1
Tel: 031-229 9292

Glen Urquhart Road,
Inverness
Tel: 0463-702468

Trade Associations:

British Chemical Dampcourse Association

16A Whitechurch Road, Pangbourne,
Reading,
Berks RG8 7BP
Tel: 07357-3799

British Wood Preserving and Damp-proofing Association

Building No.6, The Office Village,
4 Romford Road, Stratford,
London E15 4EA   Tel: 081-519 2588

Building Employers Confederation

82 New Cavendish Street, London W1M
8AD Tel: 071-580 5588.

CORGI Confederation for the Registration of Gas Installers

4 Marine Drive, Edinburgh.
Tel: 031-552 6960.

Electrical Contractors Association of Scotland

23 Heriot Row,
Edinburgh EH3 6EW
Tel: 031-445 5577.

Federation of Master Builders

Gordon Fisher House,
14 Great James Street,
London WC1N 3DP
Tel: 071-242 7583

Guarantee Protection Trust

PO Box 77, 27 London Road,
High Wycombe,
Bucks HP11 1BW
Tel: 0494-447049.

Heating and Ventilating Contractors' Association

Bush House, Bush Estate, Penecuik,
Midlothian
Tel: 031-445 5580.

Institute of Plumbing

64 Station Lane, Hornchurch,
Essex RM12 6NB
Tel: 0402-72791.

Lead Contractors Association

31 Marsh Road, Pinner,
Middlesex HA5 5NL
Tel: 081-429 0628.

National Federation of Roofing Contractors

Scottish Regional Association,
13 Woodside Crescent, Glasgow.
Tel: 041-332 7144.

National Inspection Council for Electrical Installation Contracting (NICEIC)

Alembic House, 93 Albert Embankment,
London SE1 7TB
Tel: 071-582 7746.

Scottish Building Employers Federation

13 Woodside Crescent,
Glasgow G3
Tel: 041-332 7144.

Scottish Glass Association

13 Woodside Crescent
Glasgow G3
Tel: 041-332 7144.

Scottish Master Wrights & Builders Association

26 West Nile Street,
Glasgow G2
Tel: 041-221 0011.

Scottish Master Slaters & Rooftilers Association

13 Woodside Crescent,
Glasgow.
Tel: 041-332 7144.

Scottish and N. Ireland Plumbing Federation. (SNIPEF)

2 Walker Street,
Edinburgh EH3 7CB
Tel: 031-225 2255.

Stone Federation

82 New Cavendish Street,
London W1M 8AD
Tel: 071-580 5588

Products and Suppliers:

Alumasc Ltd

Station Road, Burton Latimer, Kettering,
Northants NN15 5JP
Tel: 0536-722121
(cast aluminium gutters)

Thomas Ashworth & Co Ltd

Sycamore Avenue, Burnley, Lancs
(Speedy moisture meter)

Avongard Products Ltd

61 Down Road, Portishead,
Bristol
Tel: 0272-849782
(tell-tales for crack monitoring)

British Gypsum Ltd

East Leake, Loughborough,
Leics LE12 6HX
Tel:0602-214161
(plaster and plasterboard products)

Burlington Slate Ltd

Cavendish House, Kirkby-in-Furness,
Cumbria LA17 7UN
Tel: 0229-89661

Don and Low Ltd

St.James Road, Forfar,
Angus DD8 2AL
(makers of geotextile felt for lead underlay)

Dorma Door Controls

Dorma Trading Park, Staffa Road,
London E10 7QX
Tel: 081-558 8411
(door closers)

Durabella Ltd

9 Glendower Street, Brasenose Road
Industrial Estate, Bootle,
Merseyside L20 8PS
Tel:051-9335102
(resilient batten systems for soundproofing)

Glynwed Foundries

PO Box 3 Siclair Ketley,
Telford TF1 4AD
Tel: 0952-641414
(timesaver joints and cast iron pipes)

Hutton and Rostron

Environmental
Investigations Ltd
Netley House,
Gomshall, Guildford,
Surrey GU5 9QA
Tel: 048-641 3221
(Specialist rot surveys, Rothounds,
Curator system)

Keim Mineral Paints Ltd

Muckley Cross, Morville, Nr Bridgnorth,
Shropshire WV16 4RR
Tel: 074-631543
(stone restoration mortar and paints for
stone)

Linotol Products

PO Box1, 63 Black Street,
Airdrie ML6 6LY
Tel: 0236-62273 (Linostone, linotex,
special paints and stone repair mortars)

Marley Roof Tile Co Ltd

1 Suffolk Way, Sevenoaks,
Kent TN13 1YL

Neolith

Peel Mill, Market Street, Whitworth,
Nr Rochdale, Lancashire OL12 8HN
Tel: 0706-852731.
(non-caustic chemical stone cleaning)

Nuralite (UK) Ltd

Nuralite House, Canal Road, Higham,
Shorne, Rochester,
Kent ME3 7JA
Tel: 047482-3451
(bitumen type flashings)

Perkins and Powell

Cobden Works, Leopold Street,
Birmingham B12 0UJ
Tel: 021-772 2303
(brass ironmongery for sash and case
windows).

Protimeter PLC

Meter House, Fieldhouse Lane, Marlow,
Bucks SL7 1LX
Tel: 06284-72722
(dampness moisture meters and testing
facilities)

Redland Roof Tiles Ltd

Station Road, Cowie,
Stirling FK7 7BP
Tel: 0786-811791

Ruberoid Architectural

St Mungo Street, Bishopbriggs,
Glasgow G64 1QX
Tel: 041-772 1117
(patent glazing bars for rooflights).

Sloan & Davidson Ltd

Stanningly, Pudsey,
Yorks LS28 7XE
Tel: 0532-571892
(cast iron rainwater pipes, gutters and
hoppers, (many original victorian
castings).

Sovereign Chemical Industries Ltd

Park Road, Barrow-in-Furness,
Cumbria
Tel: 0229-870800 also in Glasgow at
041-647 9117 (damp-proof course
chemicals)

T-Pren available from:

D.Blake & Co.Ltd. 10 Beaverhall St,
Edinburgh EH7 4JE
Tel: 031-556 9632
(expansion joint system for lead gutters)

Velux Co Ltd

Telford Road, Eastfield Industrial Estate,
Glenrothes East, Fife
Tel: 0592-772211 (rooflights)

Excel Industries Ltd

13 Rassau Industrial Estate, Ebbw Vale,
Gwent NP3 5SD   Tel:0495 350655
(warmcell cellulose fibre insulation)

Selected Reading

Ashurst, John and Nicola, (1989)
Stone Masonry, Vol.1: Practical
Building Conservation. Gower Technical
Press.

Davey, A., Heath, Bob.,
Hodges, Desmond., Ketchin, Mandy.,
Milne, Roy. (1986),
**The Care and Conservation of
Georgian Houses**. Butterworth for
ENTCC.

Gourlay, Charles, (1903) **Elementary
Building Construction and Drawing
for Scottish Students**. Blackie and
Sons.

Himsworth, Chris, (1989) **Public Sector
Housing Law in Scotland**, The
Planning Exchange.

Hollis, Malcolm, (1986) **Surveying
Buildings**, Surveyors Publications

Horsey, Miles, (1990) **Tenements and
Towers**, The Royal Commission on the
Ancient and Historical Monuments of
Scotland.

Jackson, Albert and David Day, (1986)
Complete Do-it-yourself Manual,
Collins

King, Elspeth (1991), **People's Pictures
- The story of tiles in Glasgow**,
Glasgow Museums.

Marshall, Duncan and Derek Worthing
(1990), **The Construction of Houses**,
The Estates Gazette Ltd.

McAllister, Angus and Guthrie, T.G.
(1992), **Scottish Property Law:
An Introduction**. Butterworths.

National and Provincial Building
Society, (1991) **The N&P Guide to
House Restoration**. Kogan Page Ltd

Oliver, Alan C., **Woodworm, Dry Rot
and Rising Damp**. Sovereign Chemical
Industries Ltd

Robert Gordon Institute of Technology
(1991), **Stonecleaning in Scotland:
Research Summary.** Historic Scotland,
Scottish Enterprise and RGIT.

Scott, K.L. and Skelton, D.A. (1989),
Home Improvement Contract Kit.
Architecture Design and Technology
Press, 128 Long Acre,
London WC2E 9AN.

Scottish Consumer Council (1984),
**Under One Roof: A Guide for Flat
Owners in Scotland.** HMSO

Simpson, M.A., and Lloyd, T.A. (1977)
Middle Class Housing in Britain.
David and Charles, London.

Worsdall, Frank, (1979)
The Tenement: a Way of Life.
Chambers.

Index

Note: page numbers in *italics* refer to illustrations